SOCIALLY ENGAGED

THE AUTHOR'S GUIDE TO SOCIAL MEDIA

SOCIALLY ENGAGED

THE AUTHOR'S GUIDE TO SOCIAL MEDIA

TYRA BURTON   JANA OLIVER

Published by
Prism Books
MageSpell LLC
P.O. Box 1126
Norcross, GA 30091

**Socially Engaged:
The Author's Guide to Social Media**
ISBN: 978-1-941527-00-9
Copyright ©2014 Tyra Burton & Jana Oliver
Cover Design by Clarissa Yeo
Author Photographs by Celestial Studios

www.SocialMediaMuses.com
www.Facebook/SocialMuses
@SocialMuses

All rights reserved.
No part of this book may be reproduced or transmitted in any form or by any means now known or hereinafter invented, electronic or mechanical, including but not limited to photocopying, recording, or by an information storage and retrieval system, without the written permission of the Publisher, except where permitted by law.

Tyra's Dedication

To my mom, Dela Head Mitchell, happy 90th birthday!
You are my inspiration.
&
My Dad, Flem Goode Mitchell, Jr.
You will always be my Tiger.
I love you. I know you are bragging in heaven.

Jana's Dedication

To all those creative folks
who stare at a blank computer screen
and wonder what the hell comes next.

Acknowledgments

When you decide to write a book on social media, you are either insane or have an amazing support system. While no one has mistaken us for sane, we do have the good luck of working with, and being surrounded by wonderful people.

Our editor is the lovely Meghan Stoneburner, who had no clue the hot mess she was getting herself into when she said "yes" to our plea for help. She turned chapters around at the speed of light and all mistakes that are left are entirely ours.

Our typesetter Harold Buehl (a.k.a. Mage), put up with our fussiness and did a great job making us look good both for the print and e-book editions. You've never experienced picky until you've worked with a Virgo and a Scorpio at the same time.

Also our gratitude goes to Arran McNicol at Editing720 for proofing the galley, and to Brian Flatley and Skylar Sperin for keeping us on the straight and narrow when it comes to all things social media.

We also want to thank our fellow Georgia Romance Writers for supporting and inspiring us through the years.

Tyra thanks…

My husband, Shane, who puts up with my ADD and blonde moments, and luckily thinks crazy is the new normal. I love you, handsome—thanks for being in my corner and believing in me.

I'm lucky enough to have great sisters, Genia Mitchell and Miriam Garret, who are even better friends. My mom never fails to believe I can do anything I set my mind to—thanks Mom! Let's go find a Starbucks and play cards!

My former students—particularly Jennifer Beech, Joëlle Davis, Abby Foster, Nicole Neumeier, Skylar Sperin, Alaina Steiner, Alaina Stern and Meghan Stoneburner—for inspiring me and for enriching my life beyond measure.

Jennifer Ross, who frequently makes sure that I appear more "together" than I actually am. Thanks for putting up with the epicenter and being a better friend than I deserve. Brian Flatley, for keeping me up to date and being my brother-by-another-mother. Love you, man!

And last but far from least, my co-author, whom I met more years ago than we will admit. Little did I know then that I had found a lifelong friend and partner in this crazy thing we call Social Media Muses. I believe it is time to celebrate with some whiskey and chocolate!

Jana thanks…

Tyra Burton, a true Earth Mother and savvy marketing guru, for putting up with this Scorpio with all her meticulous schedules and such.

And a hug to Mage, for always being there. Because you're the light in my life.

Foreward

Social Media

Two words that bring joy into the hearts of extroverted socializers or fear to the technologically challenged across this globe. We don't know where you fit on that scale, but if you are reading this book and you need to tackle social media, we are here to help.

Just to complicate matters, social media is one of the *fastest changing promotional mediums* out there. We wrote this book during the summer of 2014, and we will lay you odds that something changed as soon as we went to print. That's the one constant of social media—it never stays the same.

In this case, technology is your friend. Print books can only go so far, so we will update the e-book edition of *Socially Engaged* as needed and post timely updates and news items on our website.

You will find a plethora of useful information on our website—**www.SocialMediaMuses.com**. Sometimes it is easier to demonstrate a concept rather than write about it. It's true, a picture can be worth a thousand words, and when it is—we'll have it on our site.

Be sure to watch for our blog posts about the latest trends in social media, as well as YouTube videos to help you understand more about how to utilize the platforms we discuss in the book.

As you start to read *Socially Engaged*, feel free to skip around. Surprisingly, we encourage you to do just that. We've packed the beginning sections of the book with the basics that we feel you need to know in order to effectively use social media to market yourself as an author. Once we dive down into the various platforms, choose which ones you have experience or interest in. There's even an index to help you easily locate topics, along with webpage links at the end of every chapter.

Note: links often change or vanish into the great void that is the worldwide web, so be sure to check the chapter sections on our website for updated information and links.

We've also included **Pro Tips** in most of the chapters to give you quick suggestions on how to boost your social media presence.

Not an author? If you are wondering if this book would be of value to you, we believe the answer is—Yes! Our examples and tactics are geared toward writers, but anyone who needs to use social media for promotion purposes can adapt what we suggest for their individual situation.

And stay tuned, during Summer 2015 we will release our second social media book, one especially designed for independent authors and the challenges they face. It will go more deeply into target marketing, analytics, newsletters and website construction, as well as cover additional platforms like YouTube, Instagram and Snapchat.

Until then—enjoy the wonderful world of social media!

~Tyra & Jana
August 2014

www.socialmediamuses.com
www.facebook.com/socialmuses
@socialmuses

Contents

1 Branding Yourself — 1
   Brand Equals Reader Expectations — 2
   Author vs. Book Brands — 3
   Your Author Brand — 4
   Taglines — 8
   Chapter One Endnotes/Links — 11

2 Social Media: In the Beginning — 12
   The Legal Side of Social Media — 17
   Trademarks — 21
   Word-of-Mouth Explosion — 23
   Chapter Two Endnotes/Links — 27

3 Getting Analytical — 28
   Touchpoints — 28
   Common Touchpoints — 31
   In-Person Touchpoints — 33
   Setting Objectives — 33
   Content Creation — 36
   Favorite Content Creation Sites — 37
   Chapter Three Endnotes/Links — 39

4 One Google to Rule Them All — 40
   Search Engine Optimization — 41
   Google+ — 43
   Personal vs. Business — 43
   Circles — 44
   Posting in Google+ — 44
   Hangouts — 45
   Events — 46
   Communities — 46

Why Google+ Matters ---------------------------------- 47
Chapter Four Endnotes/Links --------------------- 48

5 Influencers: It's All About Who You Know ----- 49
Klout -- 52
Beyond Klout -- 53
Chapter Five Endnotes/Links --------------------- 58

6 Facebook: Where Your Friends Are -------------- 59
Timeline -- 59
Page --- 60
The Key Elements ---------------------------------- 62
Organic vs. Paid Reach --------------------------- 67
Facebook Analytics -------------------------------- 69
Page Likes --- 70
Reach --- 71
Engagement --- 71
Visits -- 71
Posts -- 71
People -- 74
Why Facebook Matters --------------------------- 75
Chapter Six Endnotes/Links --------------------- 77

7 What's All This Tweeting About? ---------------- 78
The Purpose of Twitter --------------------------- 79
Logistics --- 80
Getting Started -------------------------------------- 83
Twitter Uses -- 84
Twitter Housekeeping ----------------------------- 86
Advertising -- 88
Show Me Those Numbers ------------------------ 89
Chapter Seven Endnotes/Links -------------------- 91

8 Pinterest: The New Wishbook ---------------------- 92
   Using Pinterest -- 95
   Popular Categories ------------------------------------- 96
   Your Books on Pinterest ----------------------------- 96
   Books You Love -- 97
   Secret Boards --- 97
   A Community Board ---------------------------------- 97
   Unique Interests -- 98
   Pinning -- 99
   Repinning -- 100
   Blogs and Pinterest ---------------------------------- 100
   Pinterest Analytics ---------------------------------- 101
   Finding Influencers on Pinterest ----------------- 102
   Tips and Tricks --------------------------------------- 103
   Chapter Eight Endnotes/Links -------------------- 105

9 Blogs: Journals of the Digital Age --------------- **106**
   Creating a Blog -------------------------------------- 107
   Tumblr --- 109
   Creating & Updating Your Blog ----------------- 109
   Content -- 112
   Spam --- 113
   Blog Tours --- 114
   Chapter Nine Endnotes/Links -------------------- 116

10 Goodreads: Social Media for Readers --------- **117**
   It's All About the Readers -------------------------- 117
   Navigating Around Goodreads --------------------- 118
   Goodreads for the Author --------------------------- 121
   Making the Most of Goodreads ------------------- 122
   Ask the Author -- 123
   The Dark Side of Goodreads ---------------------- 124
   Chapter Ten Endnotes/Links -------------------------- 127

11 Amazon's Author Central----------------------**128**
   Managing Your Books-----------------------------129
   Finding Your Amazon Readers-------------------130
   Chapter Eleven Endnotes/Links------------------132

1 Branding Yourself

I know what you're thinking.

"I just bought a book on social media. Why is the first thing in the book on branding?"

Because **Social Media** and **Author Brand** are joined at the hip.

Often that is a good thing, but there will be times when it's a hot mess. We'll tell you how to deal with that mess later, but for now let's start with the basics and the two words destined to strike terror into your heart: **Author Platform**.

Yes, we're going there.

Author Platform is a term often bandied about by the denizens of the publishing world, in particular editors and publishers. They'll want you to create one and have it established before your first book is published, because it's a solid start to building your career.

If you are an "old-timer," you may already have a platform, whether you know it or not. In that case, it may need refreshing or a reboot to give it a kickstart. For independently published authors (indies), you won't have anyone nipping at your heels to create a platform, but you need one as well.

Every writer needs an author platform. But what is it?

In a nutshell, it is how you are viewed by your target audience. Jane Friedman describes an author who has a good platform as "…someone with visibility and authority who has proven reach to a target audience."[1]

That makes editors and publishers giddy.

The tent pole of your author platform is your brand, around which you gather your network, media outlets and readers. So what is a brand?

A brand is how the world, your readers in particular, perceive you. How would they describe you? What comes to mind when they say your name?

Marketing guru Seth Godin defines brand as a "set of expectations, memories, stories and relationships."[2] Godin's definition explains why bestselling romance author Nora Roberts had to employ a pseudonym (J.D. Robb) for her futuristic police procedural mystery series.

Over a number of years, Roberts created an amazing brand within the romance genre. You know when you pick up a Nora Roberts novel you'll enjoy an emotionally satisfying story about a man and woman falling in love. There's always a happy ending, and oftentimes a bit of magic thrown in for good measure.

That's your expectation, and Roberts always delivers. Those same expectations wouldn't relate well to a police procedural/mystery. Because of that, Roberts employed a pseudonym so she could pen her mysteries and readers could enjoy them without damaging her Nora Roberts brand.

Brand Equals Reader Expectations

The memories and stories your readers connect to you, and your books, are part of your brand. The relationships you build with those readers are also a part of your brand. In the past, those relationships may have been created and fostered through letters or book signings. Back then, few readers were able to have any kind of timely interaction with their favorite authors. Fans were kept at a distance, and there was a clearly defined "moat" around the author's private life and his/her interaction with the readers. The author's books *were* their brand.

Social media changed that. Now you are able to connect more intimately with your fans, and they with you. Social media is the means to do this. It can strengthen your brand, and therefore your author platform. Or do damage to both. We'll talk more about that down the line.

Author vs. Book Brands

When we say the name J.K. Rowling, what is the first thing that comes to mind? My guess? Harry Potter, or something connected to that amazing series.

Rowling's author brand is tied to Harry and the Hogwarts universe in such a way that the book or series brand is intimately (and most likely permanently) connected to her name. When the Potter series was complete, Ms. Rowling wrote *Casual Vacancy*, an adult story. Not surprisingly, reviews often used terms or references related to the Potter series. Her author platform and brand are so strong, so tied to Harry, that it is hard for readers, and even professional reviewers, to see the boy wizard as only a *part* of Rowling, the author.

It's because of this she wrote her next book under the pseudonym Robert Galbraith, to avoid the "Rowling" brand expectations. Unfortunately, without her permission, her pseudonym was revealed to the world and, once again, readers weighed in on her classic British mystery novel in Harry Potter terms.

Hopefully, over time, as the Galbraith books continue (the second of which was released in June 2014), her author platform will evolve to include both.

So, as an author, both you and your books each have a brand. These two things are tied together in an "until death do you part" marriage without any means of amicable divorce, unless you change your name and launch a whole new author brand.

However, that doesn't mean you have to invent a new persona with every series or genre change (though some authors do just that.) Consider Paul McCartney. Most of us know him as one of The Beatles and a man who writes "Silly Love Songs" or blazing rock anthems like "Live or Let Die." But he's also written symphonies such as "Liverpool Oratorio." All of these come together to create Paul McCartney's brand and his musical platform.

Judy Blume, author of the popular middle grade book *Are You There God? It's Me, Margaret,* had a rough ride when she wrote her first adult novel, *Wifey.* On her website, she writes that "some people thought I would never write another children's book, some thought I had written a real book at last, some were angry that I hadn't used a pseudonym, others that I even had such thoughts!"[3] One thing is common between her children's book and adult fiction—her humor, which is a part of her brand, regardless of your reading level.

Each new book or series should help you build your brand, but if you are tied too closely to it you may find yourself in the position of Roberts and Rowling. Which isn't a bad thing, but it means you will be living double, triple or even quadruple lives under an array of pseudonyms. Or you can just be Judy Blume.

Your Author Brand

Choosing how to brand yourself is one of the most important decisions an author must make, ranking right up there with "Do I sign with this agent?" and "Is this contract right for me?" Your brand will follow you throughout your career, so it's best to determine exactly how you want others to "view" you, both online and in person.

And here's where we get to compare an author to a Quarter Pounder. No, really. McDonald's (and the other chain restaurants) take great pains to ensure you know their brands. When you walk into one of their restaurants, you expect a certain experience, and usually they deliver. There may be regional differences between a McDonald's in Austin, Texas and one in Hong Kong, but you definitely know the drill: they've taught you how to order your meal, how the food will be presented and what the experience will entail. Consistency is the key.

Your author brand must have a similar consistency because it will help the readers know what you're all about while moving them one step closer to checking out (and purchasing) your books, if you fit their reading tastes.

You *want* your readers to forge an emotional connection with you and your work. When they see your name, they know what you're all about and you want them willing to step onto your magic carpet to take another fantastic ride. This emotional connection can be enhanced through social media.

As mentioned earlier, when someone picks up a book by Nora Roberts they know they're going to be reading a romance novel, most likely with an HEA (Happy Ever After) or an HFN (Happy For Now) ending. But the experience they'll receive with a Nora story is completely different than with author Karin Slaughter, who writes dark and gritty police procedurals where there is rarely a happy ending, but always includes a solid helping of gruesome murders and difficult investigations. Karin pulls no punches. She's visceral, in your face, and she makes sure you know it. That's her brand, just as Nora's romances are hers.

But wait, we're talking about books, not the author's personality, right? It depends. Your brand may only be based on what genre you write, though that may prove difficult over the long term. Today's authors must be able to shift genres to remain published as readers' (and editors') tastes change.

The more nimble approach is to develop a brand that encompasses your personality and the "feel" of your books. Are you a crusader for social causes and your stories reflect that? Are you an unabashed romantic and your love stories are all about finding one's soulmate? Are you athletic and your books tell soaring tales of adventure, filled with pulse-pounding danger?

It is important to narrow your focus (you can't be all things to all people) so let's take a quiz! We know, most people don't like these, but it's vital that you understand how you perceive yourself in regard to your author brand, personality and platform.

So here goes: Find *three* words that best describe you. You can do this while sipping a cup of coffee, tea or something stronger. All you need is three little words.

We'll help you out…

Jane Author sees herself as:
- Passionate
- Adventurous
- Inquisitive

While Bob Author sees himself as:
- Romantic
- Idealistic
- Funny

The brand (and associated website, blog and social media presence) for Jane Author will be completely different than for Bob Author, as they should be. The readers of Author #1 will expect rousing tales filled with danger and intrigue while Author #2's readers crave "feel-good" romances.

Choose those words carefully and then develop your brand based on those terms. Don't go too vague—like fair, kind, friendly—but use bold and descriptive terms as they'll serve you better when you're forging that brand.

But it's not all about you or your readers…

In the real world, reader expectations aren't the only ones you'll have to factor into your brand. Publishers, both small and large, are weighing in on social media, and expecting authors to shoulder their share of that burden.

Sometimes they even write those expectations into your contract, at which point they cease being "we'd like you to do this" and become "thou shalts." If everything works perfectly, how you present yourself to the world should dovetail nicely with how your publisher(s) will market you and your books. But not always, as often authors and publishing houses do not see eye to eye.

It's important to ensure that there isn't a disconnect between how your publisher portrays you to the world, and how you present yourself. Having your publisher present you as a stay-at-home mom who loves to crochet and write sweet inspirational romance novels will butt heads with your online images of your roller derby races on the weekends. Unless you happen to belong to a Christian roller derby team and are raising money for homeless charities during your bouts, that's a huge disconnect. Disconnects confuse readers, and that is never a good thing.

That isn't to say you can't live your life as you see fit, just be mindful that you are now a *public* figure, albeit far less visible than J.K. Rowling. Unless you're very well known, you can wander out to the market to buy cantaloupes in holey jeans and no one will care, or even notice. But posting, blogging and tweeting about activities that are grossly at odds with your brand can be detrimental. More on the pitfalls of social media in a bit.

If you're an indie author and run your own publishing house, then your social media realities are different. You are assuming both roles, so you must take care to ensure that your brand and social media presence are in line with your goals as publisher. Tricky, but doable. Again, avoid that disconnect. Be consistent and the readers will reward you by buying your books over and over.

Taglines

Taglines are a short, pithy sentence or two which encapsulate who you are. For example, newly published author, Anna Steffl, uses "Fall in Love with Fantasy" as her tagline.[4] She writes epic fantasies with romantic elements that mainly target female readers and her tagline perfectly explains her brand and books.

Author Heather Thurmeier's tagline is: "Heart, Humor and a Happily Ever After" and her website reflects that.[5]

How about "Always Feisty, Always Funny?" That might fit author Jennifer Crusie, who is known for her madcap romances and mysteries populated with wacky heroines and sparkling humor.[6] It certainly fits a woman who writes posts with such titles as: *A Writer Without a Publisher Is Like a Fish Without a Bicycle* (a clever play on feminist Gloria Steinem's

"A woman without a man is like a fish without a bicycle.")

Creating your tagline is going to require some heavy lifting. Trust us, the first twenty you create will probably suck, but eventually you'll weed it down to something that fits you. Try your favorite tagline (and the runners-up) on your fellow authors and friends. Rewrite as necessary. Once that tagline is yours, use it! Put it on your website, your blog, Twitter and Pinterest pages. Drop it on your business cards and at the end of your e-mails. It's an integral part of your brand and needs to be seen.

Which leads to that vital piece of cyber real estate—your website. Think of this as your landing pad. You've stirred up interest on various social media platforms and the interested parties need somewhere to go to learn more about you and your books. In this case, it's your website. Using a blog as a landing pad is an option, but it's harder to give the visitor an overview of who you are as the blog's contents change quite frequently. Or at least they should. Using Facebook is an option, but not everyone wants to be on that platform. For maximum availability, a website is best.

Your website doesn't need to be fancy or expensive, it just needs to match your brand. Think about the first impression you want to convey to your readers and ensure that's what they see. Update the content regularly. Nothing is more irritating than visiting a site to determine where an author will be signing or appearing and the appearance schedule is two years old. Or worse, not there at all.

Ensure you have a contact form on that site so readers can write you. You'll come to love those e-mails, as they are tangible evidence that people care about your writing and are keen to read your next book.

Besides echoing your brand, your website should be easy to navigate. Help the visitor find the information they need. If you write multiple series, keep book lists (in order, please) in a handy drop-down box so folks know which one to read next. Make it easy for your fans to find what they need.

To find new ideas for your website's appearance, conduct a tour of other authors' sites in the same genre. Decide what works, and doesn't work, and then apply those ideas to your own site. Above all, use your tagline and brand-related imagery and graphics to reassure the visitor that they're at the right location and that you're the author they want to read.

Now that we've discussed author brand, platform and a "landing pad," let's dig deeper into social media itself.

Chapter One Endnotes/Links

1. Jane Friedman—(http://janefriedman.com/2012/03/13/author-platform-definition/)
2. Seth Godin—(http://sethgodin.typepad.com/seths_blog/2009/12/define-brand.html)
3. Judy Blume—(http://www.judyblume.com/books/adult/wifey.php)
4. Anna Steffl—(http://www.annasteffl.com/)
5. Heather Thurmeier—(http://heatherthurmeier.com/)
6. Jennifer Crusie—(http://www.jennycrusie.com/)

2 Social Media: In the Beginning

While some of you may avoid social media like you would a rabid squirrel, others may be fully engaged in every platform available. But what exactly is social media? What makes one site social and another not?

Social media is about communication and interaction between people who share their experiences through written words, videos and photos. The conversations run the gamut from silly to heavy, with cats and soldiers taking center stage. There are ruminations on daily life, celebration of good news and commiseration with the bad news.

At its best, social media is the expression of the hopes and aspirations of humanity.

Distilling this down, social media is about multi-way Internet communications. Sometimes they are topic oriented, like at Goodreads, and other times the conversations are all over the place, like on Facebook. The reviews on Amazon and Barnes & Noble are social, but in a different way than the exchange of tweets on Twitter are social.

Though increasingly platforms are requiring some form of verification, one of the enticing things about social media is the ability to "be" anyone, or even to be anonymous. Being anonymous allows you to be judged by what you post regardless of your gender, age, physical appearance or race. Early on, this was liberating for many women, who didn't have to worry that their gender would outweigh their words.

Now, anonymity opens the door for hurtful behavior, all the while lacking real-life ramifications when the posts are taken negatively. This is the dark side of social media in what is normally a friendly, conversational landscape.

Social media is the telegraph of the twenty-first century, the ability to keep your readers in the loop, in real time. Why should authors go to all the trouble interacting with people they don't know? Because in today's publishing climate, not taking advantage of that connectivity will impact your ability to sell books and build loyal readership. Increasing book sales figures lead to future contracts, which makes connectivity a must.

Being social is the primary way to ensure your fans will know when your new books launch, when (or if) you change publishing houses and how they can purchase your backlist. You can connect with readers across the globe from the comfort of your couch or desk chair. You can build meaningful relationships while networking at the same time. Readers *love* this.

However, there is a downside to all this digital "intimacy." Social media documents our lives from cradle to grave, and as a public figure you have to decide how much of that information you wish to release to your fans. Think of your personal life as a castle with a broad moat. Which people are "inside the moat" (friends, family, fellow authors)? Which people are "outside the moat" (readers and others)?

Some authors keep a very business like image on social media. It's just about their books and the business of publishing and nothing else; their personal life is off limits. There is no mention of politics or current events (other than relating to the books), and there are sparse references to their family.

Other authors allow us to peek behind the curtain, writing about their trials and tribulations. Many share family anecdotes about their pets, their son or daughter's first tooth, potty training, or graduation from high school. Weddings, funerals and other passages of life are shared as if the social media community is an extended family.

Only you can determine how comfortable you are with revealing yourself to others. We're all different. Some people are very open, with no detail held back. Other authors run under a pseudonym because of family or job-related reasons. Need an example? Hiding your identity online is very smart when you're a kindergarten teacher who happens to write erotica on the side. Not everyone is open-minded, especially when children are involved.

Speaking of kids, authors are often leery of posting pictures of their children online. Those that do may showcase their offsprings' exploits but use code names for them, such as Thing 1 and Thing 2, Rock Star or Kid 1 and 2. A pair of authors we know call their daughter Sonic Boom, based on the animated television series. That anonymity is important for young children, as you can never be sure who is reading your posts and what their intentions are.

Some authors go as far as using a post office box in another town for their mail, and listing that address when they register their website URL so someone cannot easily find their home address. Hosting companies often offer privacy add-ons that mask the address of the registrant. Though this all seems a bit over the top, keeping your home address private is a wise move. Not all your readers may respect your idea of personal space.

Just like in fairy tales, another particularly nasty side effect of social media's bridges is that they invariably harbor **Trolls**. Trolls are a peculiar social media monster who live to "flame" other posters in an effort to stir up controversy. They really like making people angry, so a seemingly harmless conversation can degrade into a shouting match (usually in all capital letters), especially when the postings are degrading or racist. If the troll is anonymous, the venom can be devastating.

The best way to handle trolls is not to rise to their bait. They live for that, so if you starve them of oxygen and ignore them, they'll (hopefully) find somewhere else to go. At times your fans may rise to their bait and defend you against a troll's attack. If the troll retaliates against your readers—this is where you need to consider getting involved.

Bottom line: never let trolls attack your fans.

They're your "peeps" and it's up to you to guard them like the precious jewels they are. If they're trolling your personal social media space and you want them gone, either block them or report them, especially if they have broken the **Terms of Service (TOS)** of that particular social media platform.

Unfortunately, not all the damage is done by trolls. Because social media is so accessible and so pervasive, if you screw up on the Internet, it's possible for the world to hear about it.

Receive a critical review? You have three options: thank the reviewer (while gnashing your teeth), ignore the review (while still gnashing those teeth) or flame back because it's obvious the reviewer is an idiot.

The third response can go viral in a heartbeat. People will remember your name (and what you said) for a long time.

If you think no publicity is bad publicity, you're wrong. Publishers, editors, reviewers and readers all watch what you say online, and if they believe you're an insincere jerk, or not really sane, they file that information away for the next time they encounter you. Which might be when you send that shiny new manuscript or book to an editor or reviewer.

Even worse, you will be mocked online, your sales will suffer and, if you really do something over the top, you could lose your publishing contract (depending on its terms.) And remember readers' emotional connections to you and your books? Bad online behavior can break these connections and diminish their desire to buy your work.

Politics, religion and publishing (sometimes) are also lightning rods. Authors often avoid the first two at all costs; the last they broach with caution as there are landmines in that topic as well. The current land war between traditional and indie authors can be brutal at times.

If you wish to be political or controversial, the safest way is to keep your personal thoughts to your private pages with a more selected readership. However you want to do this, just beware that once it's on the Internet, it's there FOREVER and what you post can go viral in an instant.

If you do want to stray where the angels fear to tread (as author Anne Rice frequently does on her Facebook page), be willing to accept some readers may be turned off by your posts, which could cost you book sales.

If you write books about gay and lesbian couples and you post in social media about your support of gay marriage, this will likely endear you to your target market.

Those who hold the opposing view would not likely be part of your target market, and the impact on them not liking your brand would be minimal in terms of sales because they would not form a connection to your brand or author platform. No matter what, it always comes back to brand and your author platform.

Which leads to the legal side of social media, or how not to get in deep, deep trouble, the kind that requires a lawyer.

The Legal Side of Social Media

As with all things related to the Internet, there are legal ramifications to using other people's content in social media. Laws are in place to protect **Intellectual Property,** be that books, software, photos, videos, artwork, etc.

So let's start by giving you a general overview of **Copyright** and **Trademark** law (we promise this won't hurt a bit) and how it applies to your social media platforms. Posting someone else's photos or media content without their permission is just as illegal as someone uploading your book for free. Just sayin'…

First, let's define copyright from the folks who should know—the United States Copyright Office.

> **A form of protection provided by the laws of the United States for "original works of authorship," including literary, dramatic, musical, architectural, cartographic, choreographic, pantomimic, pictorial, graphic, sculptural, and audiovisual creations. "Copyright" literally means the right to copy but has come to mean that body of exclusive rights granted by law to copyright owners for protection of their work.**[1]

In plain English, you, as the creator of the work, own the rights to that work the instant it is put in a *tangible* form. Just thinking about a book, website or a piece of software for the last five years doesn't cut it. When you actually do something about that idea, then things get real. The mere act of writing a book (on a keyboard, notepad, sticky notes, on the backs of used envelopes) copyrights that material.

By registering the copyright at the Copyright Office, you increase your chances of a successful court case and the ability to sue for damages should someone violate your rights.

Filing for a copyright is not hard—it can be done online at their site for thirty-five dollars, or eighty-five dollars if you mail in your application. Sometimes your publisher does this for you, depending on the terms of your contract. Sometimes it's up to you. In about a year, you'll receive a lovely document indicating that your book, *The Secret Life of Honey Badgers*, is registered. No matter what, ensure you have protected your work, because that effort may pay off if someone rips you off and posts your work for sale without your consent. Because that happens more often than you might believe.

According to the U.S. Copyright Office, infringement occurs when "a copyrighted work is reproduced, distributed, performed, publicly displayed, or made into a derivative work without the permission of the copyright owner."

A classic case from 2006: Lori Jareo, a fan fiction writer ran afoul of George Lucas and his Star Wars copyright when she self-published *Star Wars Episode IV: A New Hope*. She claimed it was just for her and her friends, but the moment she put that book up for sale on Amazon, she violated Lucas' copyright. No surprise, George's empire has a slew of ravenous lawyers, and they descended like, well, a slew of ravenous lawyers would. The book was quickly pulled from circulation, though used copies are still available at astonishing prices.

Lucas had the right to sue Ms. Jareo for damages because she'd violated his copyright. While it could be argued that *A New Hope* had zero chance of impacting Lucas' bottom line, he protected that copyright to ensure that later incursions by others would be taken seriously.

It is important to note that not everything can be copyrighted. Book titles, for instance, are not covered by the law, and that's why there's ten dozen books on Amazon with the same title. Names may not be copyrighted and neither can ideas because you can't really "fix" them in any tangible form.

Works by the federal government (like that quote from the Copyright Office, for instance), facts, information, typefaces or books in the public domain are not subject to copyright.[2]

Be aware that copyright law can vary from country to country. What is kosher in the U.S. may not be kosher in European Union or in the United Kingdom.

U.S. copyright law has changed over the years, but currently if the materials were created on or after January 1, 1978, the copyright extends for the life of the author plus seventy years *beyond* the death of that author. In the case of joint works, the same time frame applies based on the last surviving author.

To become familiar with legalities of copyright law, the Copyright Office offers an amusing animated presentation designed for students and teachers. It's a bit simplistic, but does cover topics such as "It's on the Internet, Can I Use It?", "Can Anyone Ever Use My Work Without Permission?", and an overview of Fair Use.[3]

Why is this a big deal in social media? Because if you wish to use copyrighted content, you have to understand the term Fair Use.

Section 107 of the copyright law lays out the various situations where you may use other people's copyrighted content without their permission and those situations are quite narrow.

In essence, Fair Use boils down to:
- If the purpose is for non-profit educational use
- The type of copyrighted materials you're using

- The effect using this material will have on the original work in terms of altering its value or monetary worth
- The amount of the material you use in relation to the original work

As with most things in law, there are a lot of gray areas. That's what keeps lawyers in business.

The 1961 *Report of the Register of Copyrights on the General Revision of the U.S. Copyright Law* cites examples of activities that courts have regarded as fair use:

"… quotation of excerpts in a review or criticism for purposes of illustration or comment; quotation of short passages in a scholarly or technical work, for illustration or clarification of the author's observations; use in a parody of some of the content of the work parodied; summary of an address or article, with brief quotations, in a news report; reproduction by a library of a portion of a work to replace part of a damaged copy; reproduction by a teacher or student of a small part of a work to illustrate a lesson; reproduction of a work in legislative or judicial proceedings or reports; incidental and fortuitous reproduction, in a newsreel or broadcast, of a work located in the scene of an event being reported."

Want an easy checklist? The good folks at the Copyright Advisory Office at the Columbia University Libraries has one for you.[4]

Why did we just spend several pages talking about copyright and Fair Use? Because you are going to need content for your social media, and we would prefer you didn't end up broke or behind bars.

So where do you go to find photographs and materials that won't get you in trouble? The following are a few of our favorite resources—some free while others require payment.

- Morguefile (free photos)[5]
- FreeImages (pro-grade photos)[6]
- Coverphotoz (make Facebook profile and page images)[7]
- Recite This (creates quote pictures using various backgrounds)[8]

One ingenious way to deal with this "can I use this or not" problem is a **Creative Commons** license.[9] Creative Commons (CC) is a non-profit organization whose purpose is to help people share their copyrighted content with others using CC's free copyright licenses. This allows the owner to state that "some rights are reserved" rather than "all rights."

Creative Commons has a variety of licenses, which range from rather restrictive to allowing others to remix, tweak and share your work (commercially or non-commercially) without having to directly obtain your permission. It's a cool concept that removes a lot of the gray areas relating to copyrighted materials.

If the copyright holder does not have a Creative Commons license, ask for permission to use their work. If you love someone's music, photograph, artwork and/or written material (that includes blogs and websites), take the time to honor their copyright.

You won't always get an answer to your request, but it is proper procedure to ask, and asking helps avoid legal issues down the line. If you value your copyright, value everyone else's.

Trademarks

A Trademark is an entirely separate entity from copyright. Once a "mark" is registered with the government, you gain additional legal options in case of infringement. Unlike a copyright—which requires no further maintenance once it's registered—a trademark requires additional filings and fees to retain those rights over the years.

So what is a trademark? According to the United States Patent and Trademark Office it is:

"...a word, phrase, symbol or design, or a combination thereof, that identifies and distinguishes the source of the goods of one party from those of others."[10]

Trademarks, unlike copyrights, break down into specific classifications and have a trickier (and more expensive) application process. At the current time, filing a trademark is expensive—as high as 365 dollars *per class* for paper filings, and 50 dollars cheaper for online filings.

It is possible to DIY your trademark application, but most people use an attorney. Like a copyright, it takes up to a year for the process to be completed once you prove you have the right to register the mark.

If there are any contingencies or objections, additional materials may need to be filed to substantiate that you are entitled to register the mark. The USPTO.gov site offers a database of trademarks with a handy search engine and a guide on how to file for your mark.[11] If the mark is granted, you may add an ® after the name to indicate it is registered.

Movie studios trademark a wide range of goods under various classifications, no doubt as per terms of their contract with the authors. For instance, there are numerous trademarks for Hunger Games and Percy Jackson.

Author Sherrilyn Kenyon has trademarked her Dark-Hunter® name under the categories related to movies and television, printed materials (books, bumper stickers, notecards, etc.) and clothing (tee shirts, tank tops, hats, etc.) Jana Oliver, co-author of this book, has trademarked her Time Rovers® and Demon Trappers® names under printed materials.

But why would you go to all that trouble and expense? Much like a copyright, registering your trademark allows you to control how that is used. If someone wishes to produce Dark-Hunter tee shirts, they must contact Ms. Kenyon to receive her permission. Failure to do so may result in a Cease and Desist order and a potential lawsuit.

Also, if you have specific logos you utilize for your brand, you may want to consider getting a trademark to cover them as well.

Now that you've survived the legalese, let's shift to how social media has created a Word-of-Mouth Explosion...

Word-of-Mouth Explosion

Let's travel back in time for a moment. It's the early 1990s and you've finally run through your huge TBR (to be read) stockpile. You need a book.

You make your way to a bookstore or maybe even the library. You stare at the shelves, picking up books with interesting covers and intriguing back cover copy. You narrow your selection down and then ask a clerk or a librarian their opinion about the books you've selected.

Or maybe you mention to your best friend, who is a voracious reader, that you are without a book and she tells you all about the latest mystery she read. It's by a new author that she tried because a member of her book club raved about it at their last meeting.

These scenarios depict **Word of Mouth.**

Back in the days before the Internet, word of mouth was the way we found a new dentist, decided if we wanted to eat at a certain restaurant, and yes, how we decided what book to read next. Word of mouth was the reviews, recommendations or warnings for certain services and products that we shared with our friends and family. Sure you might read the *New York Times* review of a hot bestseller, but what really mattered was whether or not your best friend liked it.

Fast forward to the new millennium, where word of mouth is alive and well on the Internet. And it has exploded beyond what most could've imagined.

In his book *Socialnomics,* Eric Qualman refers to the transformation of Word of Mouth on the web as becoming **World** of Mouth. We are no longer limited to the opinions of those we know and who are geographically close to us. We now can read product reviews on multiple sites (from across the globe) and those insights are right at our fingertips.

A Google-published book, *Winning the Zero Moment of Truth: ZMOT* by Jim Lecinski, discusses how consumer behavior has changed because of our instant access to information. ZMOT, or the Zero Moment of Truth "…is that moment you grab your laptop, mobile phone or some other wired device and start learning about a product or service you're thinking about trying or buying."[12]

Before the Internet, we'd require some event or stimulus to kickstart a need for a product. We'd then go to the store shelf to make our decision, and lastly, after purchasing it, we would experience the product we bought.

Now the Internet shakes this process up by inserting ZMOT between the stimulus and the store shelf, and that store shelf may not be a physical one but instead be virtual, in an online store. We still have the experience with the product, but now we can tell the world what we thought of it.

So, why does this matter to you, the author? Back in the day, you and your publisher would try to get in the good graces of librarians, booksellers and professional reviewers, as they were your primary word-of-mouth outlets to potential readers. In today's digital world, they are only part of the picture.

Reviews on Amazon or Barnes & Noble can be accessed from almost anywhere at any time and can be written by anyone. Book blogs, created by individuals or reader groups, may have greater numbers of followers than many authors. To understand the power of consumer review blogs, just check out Smart Bitches, Trashy Books.[13]

Established in 2005, a group of romance readers who were "eager to talk about which romance novels rocked their worlds, and which ones made them throw the book with as much velocity as possible" has grown not only into a website, but also has spawned two books. One of them, *Beyond Heaving Bosoms: The Smart Bitches' Guide to Romance*, is used in undergraduate courses at Yale, Princeton and DePaul University, and the site has appeared in *USA Today*, *The New Yorker* and the *New York Times*.

The new world of mouth doesn't stop there. Social media sites like Goodreads and Shelfari encourage readers to not only review books, but to share what they are reading, and want to read in the future. Readers are having multi-way conversations with "marketers, friends, strangers, websites and experts," all of which compete for their attention.[14]

We started off this chapter defining social media, and now that we have traversed the "exciting" landscape of copyright and trademark and made our way to ZMOT, we hope you have a better understanding of the impact of social media. Your content in social media, as well as on your website, may be your readers' first experience with your brand. Or maybe it will further enhance the emotional connection they have formed with you by reading your books.

Regardless, your online world helps define your brand and author platform. It is the primary place for readers to have ZMOT moments, and to become acquainted with the author behind the book.

Chapter Two Endnotes/Links

1. Copyright Office—(http://www.copyright.gov/)
2. Plagiarism Today—(http://www.plagiarismtoday.com/2010/01/08/5-things-that-cant-be-copyrighted/)
3. Animated Presentation on Copyright—(http://www.loc.gov/teachers/copyrightmystery/?#/copyright/)
4. Columbia University Libraries' copyright PDF checklist (http://copyright.columbia.edu/copyright/files/2009/10/fairusechecklist.pdf)
5. Morguefile—(http://www.morguefile.com/)
6. FreeImages—(http://www.freeimages.com)
7. Coverphotoz—(http://coverphotoz.com/)
8. ReciteThis—(http://recitethis.com/)
9. Creative Commons—(http://creativecommons.org/)
10. United States Patent & Trademark Office—(http://www.uspto.gov)
11. How to File for a Trademark—(http://www.uspto.gov/trademarks/process/index.jsp)
12. *Winning the Zero Moment of Truth: ZMOT* by Jim Lecinski (Amazon Digital Services, Inc. 2011)—(http://www.thinkwithgoogle.com/collections/zero-moment-truth.html)
13. Smart Bitches, Trashy Books—(http://www.smartbitchestrashybooks.com/)
14. *Winning the Zero Moment of Truth: ZMOT* by Jim Lecinski

3 Getting Analytical

We feel we should warn you about this chapter. It may have a technical term or two, and we're even going to talk about business practices like "objectives." But fear not! By the end of the chapter we'll be back to great tips for creating content. So, fasten your seatbelts and let's get analytical.

Touchpoints

When you say the word **Touchpoint** to a romance writer, she most likely conjures up love scenes, maybe a kiss on the back of the neck or a hand sweeping gently down a lover's body. Guess what? We're not talking about *those* types of touchpoints. Instead we're referring to anywhere or anytime you touch the reader. No, wait. That sounds like grounds for inappropriate writer/reader contact. Let's try that again.

> **A Touchpoint is any time you come in contact with a reader, be it virtually, physically, mentally or emotionally.**

If your books are available at Barnes & Noble, one touchpoint would be the store shelf where it is displayed. Or let's say someone searches for your book on Amazon.com—the webpage dedicated to it is another touchpoint. The book signing you did at the independent bookstore last week—another touchpoint. The guest blogging you did yesterday is yet another.

Touchpoints are all over social media. Your Facebook page, Twitter stream and Pinterest pins are potential places for you to come in contact with readers. The moment when a reader sits down on the commuter train and pops open your book on his e-reader—you guessed it, another touchpoint. The reviews on sites, whether retail, professional or consumer, are additional places people come in contact with your work, as are libraries, book clubs, friends, etc. An author's touchpoints are numerous, but the real question is which are the most important. Which reader encounters are key to motivating them into buying your book? Because at the end of the day, that is what you need to have happen.

Before we go further, we need to talk about the concept of a **Target Market**. As the business of writing has been pushed more on the author, the ability to clearly define your target market—the group of readers that you feel are most likely to buy your work—is vital to success. It's also the key to determining your touchpoints and appropriate social media platforms.

By knowing where the most important touchpoints are for your target market, you can attempt to "manage" their experience. The main goal is that at the critical touchpoints, you are meeting their expectations about you as an author and brand. Remember the author platform that we talked about earlier? It's what helps build reader expectations.

No matter what, you must be aware of how your publisher perceives your brand so you don't inadvertently scuttle their marketing plans. Who do they perceive as your target audience? Does it match your target market? If your publishing house says that your Young Adult romance's target audience is only teen girls between the ages of fourteen to eighteen, that group includes just one section of your total audience. In this case, they'd be missing a large section of the adult crossover market, estimated to be 55% of sales by *Publishers Weekly*.[1] If you find yourself at odds with your editor, marketing or publicity departments, then talk with them and try to find common ground. If you are unable to do so, be sure you market to those potential readers they've missed.

There are some common touchpoints for authors and they are connected to places where you establish your author platform and brand. Your website is the "hub" of your brand, the one place on the web that you control. You decide exactly what is presented to readers, and it is here that you can guide their expectations.

Visit Sherrilyn Kenyon's website, and even if you haven't read any of her books, you'll get an idea about her brand and begin to form impressions.[2] The website is in black and gray, with even the social media icons at the bottom of the page done in gray scale instead of the usual colors. The site sets the expectation that she writes "darker" books and that they aren't light and happy tales.

Now, let's look at Debbie Macomber's website, which features a bright white background with a pleasant green header.[3] This site tells readers to expect sunny, heartwarming tales and that's what Debbie delivers.

Jana's site for her Demon Trappers series sets the expectation that bad things happen to good people.[4] Her website, with the ominous red and black format, forewarns the reader that this is not the land of Chick Lit or Cozy Mystery/Romances.

It's worth repeating again—your website is the one touchpoint on the web that you control. Everywhere else, someone can dictate, modify, embellish or own the content about you. Your website is yours and yours alone. Guard it. Use it. Update it. It should be the main portal for your fans and future readers to find out more about you. With the realization that your website is the one thing you can spruce up and decorate the way you want, we must also admit, it most likely will not be the first place a reader comes in contact with you. In other words, it is unlikely that their first touchpoint will be at your website.

How is a reader most likely to find you for the first time? Remember those three magic words from a previous chapter? *Word of mouth.* It is the most likely form of communication for a new reader to discover you.

Common Touchpoints

In regard to word of mouth on the web, the common touchpoints for authors are Goodreads (see Chapter 10), Amazon, Barnes & Noble, or a blog review site. The first three are all contributed content. Barnes & Noble has a "Meet the Author" section below the description of the book and above the reviews. Amazon offers an author page (see Chapter 11), where you can provide information about yourself and your books, and provide a link to your website.

On Goodreads, a social media site designed for readers, there is the option for an author page, and you can participate as a contributor, adding reviews on the books you've read. Goodreads is filled with people who enjoy books and is a breeding ground for word of mouth. Even if you do not offer book reviews, you will want to have a presence there to monitor that your books are being correctly represented and added to the database.

Last are the touchpoints that occur on social media platforms. Even though someone else owns these platforms, you can still, to some extent, control your branding in these settings. For Facebook, this means you need an engaging cover and profile pictures. On Twitter, it means you are no longer rocking the egg avatar (see Chapter 7) and have set your background to an image or graphic that is all about you. Where should your branding motifs for these platforms come from? You guessed it—your website. It is vital that your brand be consistent across the Internet to enhance user experience with these touchpoints. By doing this, you will create the same expectations across each of these platforms for both your current and your potential readers.

> **Pro Tip: Where is your target market online? Once you figure out which platform they prefer for social media, focus on that touchpoint.**

Book covers and back cover copy are key touchpoints for authors. Eye-catching cover art draws attention, whether it's on a bookstore shelf or displayed on a computer screen. If you are traditionally published, you may have little to no say in either the cover design or the back cover copy. If you are indie published, your goal should be to create both front and back covers that set up the expectations about your story, what it is about and the quality of your work. If you can't tell the difference between an indie book cover and a New York-created one, the reader won't be able to either. Indie authors, please, spend time and some money, if necessary, to have professionally designed covers. Nothing says amateur more than a bad cover or stilted back cover copy. By putting forth a top-quality image, a good cover will tell the reader that you are a professional and serious about your work.

Remember the gentleman on the commuter train using his e-reader? He has already chosen to experience that author's work and is in the midst of a deeper touchpoint, the one where he is eager to immerse himself in the author's fictional world. There are several things that can jar a reader out of that experience: bad grammar, misspellings and plot points that leave you hanging, can all negatively influence this reader/story touchpoint. The key to this touchpoint is to ensure your work is polished before the reader ever lays their hands on your book. By excelling at the story and its presentation, you encourage the reader to invest in you again.

In-Person Touchpoints

Now, let's talk about in-person touchpoints, those rare times when you come in direct contact with your readers. Maybe it is at a convention like DragonCon, or a book festival, or a signing at the local bookstore.[5]

Regardless, this is your time to shine and really convey your brand. Jana frequently wears black, which fits with her gritty, imperfect world settings. Tyra, on the other hand, wears pinks and bright colors, fitting with her Young Adult Fantasy and Cozy Mysteries. Can you imagine a cozy mystery author showing up for a signing all dressed in black with knee-high buckle boots? It just wouldn't fit expectations. Now, if she wrote Urban Fantasy, the expectations would fit the presentation.

As a final note, sometimes it is a good thing not to meet expectations. Just make sure you are choosing to show your readers something unique to your brand, and not let them down.

Setting Objectives

Just like many of us have word count or writing time goals, objectives for marketing, and particularly for social media, will let you know if the time you are spending is helping you to accomplish your goals. The end goal for most authors is to sell books, but there are also interim goals that can help you get there.

If you are a new author, getting people aware of your brand is the first step. For established authors, you may want to expand into a different market, drive your target market to your website or have them sign up for your monthly newsletter.

One of the drawbacks many people associate with social media is the inability to measure success. This simply is not true. The ability to measure your success is linked directly to establishing appropriate goals or objectives.

When you approach social media and the rest of your marketing efforts from the perspective of goals and what you need to do to achieve them, you'll be able to measure how those efforts are impacting your bottom line, both financially and in time spent.

Your objectives should be **SMART**. By smart, we don't mean that they need to be written with ten dollar words but instead that they should meet five standards:

Specific
Measurable
Attainable
Realistic
Timed

Example 1:
Sally wants to increase traffic to her website by 20% between July 1st and October 31st. She chose this as her objective because she has a new book coming out in late July and

will be making appearances at conventions and book signings through October. Her website will have information about the latest book, sneak peeks, a contest and links to her blog. She believes by driving people to her website, they will learn more about her and her backlist, and will be more inclined to meet her (and ultimately buy her books.)

Her objectives are specific:
- Increase traffic to website by 20%
- Measurable (Google Analytics will collect data about her website traffic)
- Attainable and Realistic (can be determined by looking at traffic patterns to her site)
- Timed (she has designated four months during which to reach her goal).

Sally is using inbound marketing to increase traffic to her website by engaging people in a variety of social media platforms. She is directing her target market to arrive at a certain website or *take a specific action*. Social media platforms are great drivers for inbound marketing. Let's look at another objective for Sally.

Example 2:

Sally wants to increase Facebook traffic to her site by 10% from July 1st to October 31st. She plans to post excerpts from her new book, along with deleted scenes. She'll post teasers with a link to her website to read more. She can measure success of these posts by looking at the number of likes, shares and comments she receives on each of them, and she could choose to have objectives for each of these engagement actions.

Setting objectives is a great way to know if your social media plan is achieving what you need. It gives you a measuring stick and a way to know what to adjust to help you achieve your goals. As we go through each social media platform, we will be looking at the analytics (statistics that help you measure your success).

Content Creation

The content you use on social media is one of the keys to success. To say "content is king" is an understatement in social media. Content encourages engagement and drives people to your platforms.

Unique content is the best way to engage fans and create an emotional connection with them. Whether it's photos, contests or deleted scenes, finding the content your fans will rave about is key.

For each social platform that we cover, we will talk about what is the right content for that particular social media site. What works on Facebook doesn't necessarily work on Twitter—and in our opinion, sharing the same content on both platforms is a no-no about 90% of the time.

By setting objectives and checking each platform's analytics, you will be able to discover what kind of content is king for you. And don't worry, for each platform, we'll walk you through how to use the analytics to your best advantage.

There is one particular type of content that deserves a special mention—**Evergreen**. Evergreen content isn't time sensitive. It isn't just good for today, but can be reused. Maybe each year you write a series of "thankful" blog posts connected to Thanksgiving. When next year rolls around, you can share the previous year's blog entries on your social media channels.

We suggest you create a way of keeping track of evergreen content. A simple spreadsheet with a tab for holiday/time-specific posts and a second tab for any time evergreen content would work well. Include columns for the name of the content, where it is located on your hard drive or in a cloud (because nothing is more annoying that knowing you have this cool content but can't find it) and when it was last shared. We'll even help you out—look on our website for a template to keep track of your evergreen content.

Favorite Content Creation Sites

We both have go-to sites for creating content for use on the web, so it's only fair we share those with you.

Morguefile is perhaps Tyra's favorite site for finding inspirational content.[6] Morguefile has an extensive and searchable collection of copyright-free photos which range from totally amateur to professional grade. We have found great photos for both website and social media use here. You may want to create a folder of these images and browse through your favorites when you need inspiration.

Next up is Pixlr.com, where you can edit photos and create collages for free.[7] You just upload your photos onto their website and create photo content. You can create cover photos for Facebook, backgrounds for Twitter as well as images for your blog.

Pinwords.com is another site that helps you create content—particularly when you want to post a quote.[8] Our favorite part is that you can upload your own photo and add wording.

For those with smartphones, an on-the-go content creation tool is Over, an app available on iOS and Android platforms.[9] You can take a photo from your phone and add text to make your own memes, then easily save them or post directly online.

Google Alerts, our final tool, is not used to generate content, per se, but helps you locate content you may wish to share—particularly articles, blogs and reviews about your work.

Google Alerts is basically a search engine that sends you results when new items connected to your search appear on the web.[10] For instance, you can set up a Google Alert for your book title, and any time someone posts something about your book on the web, you will be sent an e-mail to let you know. It's an amazingly simple and handy tool.

We've covered a lot of ground in this chapter and have started the conversation about some scary topics—objectives and analytics. Hopefully, now you have a better idea about the key points on the web for attracting new readers from your target market, how to determine your objectives and how to keep your current readers engaged.

Chapter Three Endnotes/Links

1. Publisher's Weekly—(http://www.publishersweekly.com/pw/by-topic/childrens/childrens-industry-news/article/53937-new-study-55-of-ya-books-bought-by-adults.html)
2. Sherrilyn Kenyon—(http://www.sherrilynkenyon.com/)
3. Debbie Macomber—(http://www.debbiemacomber.com/)
4. Jana Oliver—(http://www.DemonTrappers.com)
5. DragonCon—(http://www.Dragoncon.org)
6. Morguefile—(http://www.morguefile.com)
7. Pixlr.com—(http://pixlr.com/)
8. Pinwords—(http://www.pinwords.com/)
9. Over—(http://madewithover.com/)
10. Google Alerts—(http://www.google.com/alerts)

4 One Google to Rule Them All

If you are on the Internet, you probably have some familiarity with Google. Its search engine is used more than any other, and between Gmail, Blogger, Picasa/Photos, Google Drive, YouTube, Google Alerts, Google+ and more, they have a substantial online presence.

As designed, Google+ (G+) was meant to be bring all these things together in one place, augmented with a social component. At the SXSW (South by Southwest) conference in 2012, a company executive proclaimed "Google+ is Google."[1]

From that moment on, they made everything in their user tool stable connected to G+. Want to comment on a YouTube video? You must sign up for a G+ account, which, of course, links to your Gmail account. Same thing for Blogger and Picasa. All in an effort to have us consolidate (and use) Google for more important online actions.

However, at the most recent Google I/O, the invite-only developers' conference, out of eighty possible sessions on Google topics, none were devoted to G+. This, combined with the lack of celebration as it crossed into its third year, and the departure of its biggest champion, Vic Gundotra, from the company, has left G+'s future a bit up in the air. But with over three hundred million registered users, one has to assume that Google will continue to place some emphasis on that platform.

As we were going to press on this book, Google+ has announced that users can now create profiles with fake names. (Up until July 2014, you had to use your real name as your G+ profile.) They've finally recognized that some people wish to remain anonymous, such an as author with a pseudonym.

It also means your book characters can now have their own G+ accounts, which opens up a whole new avenue to interact with your readers.

Because they keeping changing things, you will need to stay abreast of what is happening with G+. We will do our best with our website and social media posts to let you in on updates about this platform.

Search Engine Optimization

Search Engine Optimization (or **SEO**) is the bane of many a website owner. Your position on a search engine results page is determined by how relevant your website is to a particular keyword search. Being located on the first page of the results listing will give you a tremendous boost when it comes to connecting with new and existing readers.

When it comes to online searches, Google holds over 67% of the market share, with Microsoft/Bing and Yahoo! coming in a distant second and third.[2] So, having a feel for how Google ranks websites is vital for a socially interactive author and will help drive readers to your website.

SEO refers to the practice of ensuring your website and other key elements are fine-tuned to direct people to your home page during their keyword searches. While the formula that Google uses to determine your ranking is unknown, it's extremely complicated (more so than the one Facebook uses for its newsfeed). However, we do have a few educated guesses.

To improve your ranking, three elements are important:
- High-quality content
- A clean site, free of broken links
- No duplicated content

Here is where your writing skills can really shine, by carefully choosing both the copy and the wording on your website. Make sure to use keywords that are associated with your books and your genre(s) to help the search engines know what you are all about.

In the early days of SEO, links were the most important element, but now they are downplayed unless the links directing people to your webpage come from high-quality and respected sites. Only then will you get a ranking boost. A site that is frequently updated will be higher ranked than one that is static or rarely changes. Lastly, having a responsive website that recognizes and changes based on the device accessing it will also help boost your SEO.

Your presence in social media may also improve your SEO, particularly on G+. It is, after all, Google's social media site, so it isn't surprising that they give more emphasis to it than Facebook (whom they compete with for advertising dollars and uses Bing search integration on its site).

To understand why SEO is important, you need to recognize that Google wants to bring its search users the best results possible, so their formula relies on site cues that reflect respected content.

These cues (which we mentioned above) are what Google relies on to determine what will keep their searchers happy (and coming back).

We suggest that after you have spruced up your site, you register it with Google and Bing so their search engines are guaranteed to drop by and see what your website has to offer.[3]

Update your content regularly; keep the site clean and fresh and your rankings will improve over time.

Google+

If the above section on SEO made your head spin, fear not. If SEO was a recipe, all you'd need to do is mix together a generous helping of engaging website content and add a dash of G+ activity for good measure.

When G+ was released, it was marketed as a replacement for Facebook, and while many embraced this change, it has not been an overwhelming success. With over five hundred million users, many people have a G+ account but most are spending very little time on the site (5.5 minutes per a month versus Facebook at 6.5 *hours*).[4]

Still, G+ has many components that are author friendly, so let's dive in and see what this platform holds in store for you.

Personal vs. Business

Much like Facebook, G+ offers you a choice between a personal profile and one for a business/organization. We suggest that you just have a personal profile, because there is no cap to the number of followers you can have. You can upload a cover photo, a GIF (an image that involves movement) or a standard avatar photo. Google will then ask you for tons of personal information, ranging from where you live to what you do for a living. Make sure to check your privacy settings so that information is only available to those you want to see it.

To examine your profile, you can click on your portrait on the upper right-hand corner and hit "View Profile." From here you will also be able to edit and create posts. At this time, there isn't a social management tool that allows you to schedule G+ posts on a personal page; however you can schedule on a business page through several services, including Hootsuite (which we discuss in Chapter 7).

Circles

Google+'s architecture is based around placing friends, family and people you want to follow within designated "Circles." This allows you to choose which circles see your posts. For instance, if you're posting pictures of your new goddaughter, which you may not want to make public, you can tag it so only people in your "family" circle are able to see the photos.

You can have as many circles as you want, and a person can belong to more than one circle. For instance, maybe your Uncle Bob is in your family circle but he is also a photography enthusiast—so you have him in your photography circle as well.

Circles can be very handy if many of your readers are on G+, because you can create circles for different genre fans or series. You can even add a circle for your street team (fans that help spread the word about your books) so only they can see the content. The primary advantage of this visually driven organizational tool is that you can easily drag and drop people into different circles.

Posting in Google+

Because of the circles, it is possible to have only one G+ personal account, with individual circles designated for your family, friends and fans. Just make sure that when you post, you have indicated the correct circle before you hit "Send." If you want to tag someone in a post, simply type "+their name" and choose the correct person from the list. For some reason, posts on G+ tend to have better grammar and are more structured than on other social media.

Another way to engage on G+ is to comment on other people's posts. If you find influencers, comment on their postings. (See Chapter 5 for information on these very important people.) Anywhere you comment, people can mouse over and a "hovercard" will pop up with information about you.

You can also utilize hashtags (the # symbol) to indicate the topic of your post. This is especially helpful if you are referencing a cultural event, a world happening or if you want to indicate your book. For instance, Jana could use #DemonTrappers when talking about her Demon Trappers series. If she was to post about her book *Briar Rose*, which is a modern-day retelling of *Sleeping Beauty*, she might want to use the hashtags #SleepingBeauty and #FairyTales. Notice that within a hashtag you don't put spaces between words, and initial capital letters make them easier to read.

As with most social media, visuals can draw people's attention, so use them when you can on G+. Also, don't forget to link your G+ profile to your blog and other content on your website, the best way to help boost your SEO using G+.

Hangouts

Hangouts are Google's answer to video conferencing. Through the video chat function, you can gather up to ten of your closest friends or fans, and talk, regardless of where you live. If you have more than ten, consider forgoing the video and doing a text chat with up to a hundred people.

Another feature is Hangouts on Air. These are special Hangouts that allow you to record part of your video conversation and post it to YouTube. You are able to do some video editing and determine how the camera switches from person to person as they talk. To find out more about how to set up a Google Hangout, visit their site for details.[5]

Hangouts offer the opportunity to connect face to face in the virtual world and extend personal touchpoints to readers who may never get to meet you in person.

Hangouts can also be effective for coordinating your street team and meeting with bloggers who wish to interview (or write about) you as well as helping generate content for YouTube.

Events

You can plan, then invite people to your Hangouts (and other happenings) in Google+'s Event section. It's easy to use: When you are looking at your newsfeed, click on the "Home" button in the upper left corner to view the G+ dropdown menu. There you'll find Hangouts, as well as a tab for Events.

Under Events, you can invite the public or just certain individuals/circles. If you are doing a book signing, you may want to create an event to announce it to the world. You can also invite people to Hangouts with an event invite.

Communities

Another highlight of G+ is its Communities, which you'll find by clicking on the "Home" button in the upper left corner of your newsfeed. Once there, you can search for communities with specific topics of interests or look at G+'s recommendations.

If you search for "Urban Fantasy" you'll see community results such as Urban Fantasy Authors, Urban Fantasy Readers, Fans of Urban Fantasy and Independent Urban

Fantasy Writers. Communities can be great places to find other authors (to trade tips and tricks) and a great way to connect with readers.

Why Google+ Matters

Google+ offers a number of useful ways to connect with potential readers. Where Facebook is primarily for friends and family—though potential readers may be discovered there—G+ helps you find new people, some of whom will be readers who have never heard of you previously. Though not as pervasive as Facebook, G+ still offers benefits for the social media savvy author.

Chapter Four Endnotes/Links

1. Google+ Announcement—
(http://www.webpronews.com/vic-gundotra-talks-google-at-sxsw-draws-some-criticism-2012-03)
2. Statista Ranking for Search Engines—
(http://www.statista.com/statistics/267161/market-share-of-search-engines-in-the-united-states/)
3. Google Website Registration—
(http://www.google.com/submityourcontent/website-owner/)
Bing Website Registration—
(http://www.bing.com/toolbox/submit-site-url)
4. Google Usage Statistics Time.com—
(http://business.time.com/2013/10/03/google-is-far-from-losing-the-war-over-social/)
5. Google Hangout—
(https://support.google.com/hangouts/answer/3111943?hl=en)
(https://support.google.com/plus/answer/2553119?hl=en)
6. Google Authorship—(https://plus.google.com/authorship)

5 Influencers: It's All About Who You Know

Looking for people to spread the word about your new novel? Hoping to stumble upon a person who communicates directly with your target readership? Anyone can share reviews, opinions and expertise if they have access to the World Wide Web, but not all of them are equal in terms of "pull." You know this—but how do you find the ones that matter? This is where influencers come into play.

Influencers are individuals who have some type of impact on consumers online. For authors, influencers are people who sway readers—specifically your readers or target market. Influencers may have a large online or social media footprint, or be connected to others that do. It's all about quantity and quality, or the tradeoff between the two. Someone who has great engagement with a small number of your target readers may be more valuable to you than someone with thousands of followers that aren't part of your audience.

Influencers are everyday people, not celebrities, and for authors, influencers are people who are willing to spread the word about your books. These influencers have "leverage." A librarian in your local community who is the go-to person for Young Adult or New Adult book recommendations is an influencer with those teens. The same librarian may also be an influencer for *parents* of young adults, letting them know what story elements are in a particular book or suggesting appropriate age/experience levels.

Bookstore employees, especially at independent shops, will hand sell books they enjoy. By connecting with these in-person influencers, you can forge links to your target audience, and if they have an online presence—even better.

The reviewers on the Smart Bitches, Trashy Books site are influencers for romance books.[1] Looking for someone to review your science fiction saga? Antony Jones, the main reviewer at SFBook.com, may be a key influencer for your book.[2] For those that write speculative fiction, The Book Smugglers may be the right review site for you to contact.[3]

A champion for your particular genre may be difficult to find, but don't forget to think outside the narrow box of book influencers. It's all about finding people that have access to your target audience and can sway their buying decisions. If you write Steampunk, bands such as Frenchy and the Punk, Abney Park or The Extraordinary Contraptions could offer inroads into the Steampunk community.[4]

Also, look at genre-specific conventions like the Steampunk World's Fair[5] to make connections. Find key influencers in the greater community and they may be able to help you connect with potential readers.

If you write horror novels, then consider individuals who have prominence anywhere within the genre. Maybe you can connect to someone who has a large following in social media that loves classic horror films. Your retro horror book may be the perfect read for them.

Brand Advocate is yet another term used to label people who have an impact on their followers and friends. They love the product and would find it difficult to stop using it. Tyra is a brand advocate for Starbucks. She reposts their material throughout social media, creates posts when they have specials and patronizes one in every city she visits. Starbucks isn't just coffee to her; it is her comfort drink, her favorite place to write outside of her sunroom. She has an *emotional* connection to the brand.

Our friend Jennifer loves author Nalini Singh and has re-read her Psy-Changeling series about twenty times.[6] When a new book is released, she reads the series from the beginning to immerse herself in Singh's world again before reading the latest offering. Singh's writing strikes an emotional chord within Jennifer, something many authors strive for with their readers. She encourages her friends to try the Psy-Changeling series, and even goes so far as to suggest Nalini to strangers if she finds herself in a conversation about authors. She is Singh's enthusiastic brand advocate.

Jennifer is also an influencer on the web. She has a following on Facebook and has often posted about her love of Singh's books. Several of her friends have started reading Singh because of those posts. This is where magic happens—when a brand advocate is also an influencer.

In your case, think of your fans who are always the first to support you and those who leave reviews at Goodreads or on retail sites. They are most likely your brand advocates, and you should return the love.

When you have an advocate for your books who is also an influencer, you have won the jackpot. With a little encouragement, they will use their social presence to help you—and be thrilled to help out. Members of your street team(s) may fall into this category, if they have a presence online.

Your goal is to make these readers/fans feel special and have them want to share the good feelings they get from reading your books and interacting with you. Finding ways to give them "exclusive access" to extra content before it is released to the masses or giving them an ARC (advance reader copy) of your upcoming books are two ways to make them know they're deeply appreciated.

When Jana travels, particularly abroad, she tries to set up informal meet-and-greets with fans at coffee shops or pubs. This is a great way for her to have quality, small group time with her brand advocates. Many of these fans also have an online presence. Sister Spooky, an avid blogger and fan, attends meet-and-greets whenever she can, interviewing authors and then posts about the meet-ups.[7]

The next best thing to meeting in person is social interaction on the web. Google+'s Hangouts on Air are a great way to communicate with fans using video, but so is the simple response to a comment or question on Facebook or Twitter. (We'll be talking more about those platforms in later chapters.) Repinning a fan's photo with you from a book signing on Pinterest means a lot to that reader. The simple act of acknowledging someone's efforts, and listening to them, is not only polite, but it will score you bonus points with your readers.

Klout

One of the ways you can measure someone's online footprint is through Klout.[8] Klout is a website that measures the reach and impact of a person's social media presence. Influence, according to Klout, is "the ability to drive action." When you share something on social media or in real life and people respond, the more influential you are and the higher your Klout score.[9]

Klout rates people on a scale from 1-100 based on their actions, interactions and engagements across more than eight different social networks. The more engaged a person's following is, the higher their Klout score, even if they have a smaller following. If the Klout score is around 40, then they have average impact on the web. But if their score is above 63 they are in the top 5% of influencers.[10]

Tyra's Klout score hovers between 64-66 most days, making her a top influencer, primarily on Facebook and Twitter, with frequent posts about pets, and, you guessed it—Starbucks. Because of her high Klout score, she is both a brand advocate and influencer for the coffee house's brand. Jana's Klout score is 59, which means she's slowly reaching toward that top 5%.

Once you register for Klout, you are able to search for users with similar interests. Remember The Book Smugglers as an influencer for speculative fiction? When we did a search on "Books," they popped up with a Klout score of 62. The ladies who started the blog (Ana Grilo and Thea James) as an outlet for their reviews of "speculative and genre fictions for all ages," have over seventeen thousand followers on Twitter and were nominated for the 2014 Hugo Award for Best Fanzine.[111]

According to their site, *The Book Smugglers* are willing to receive solicited advanced reading copies and prefer Speculative Fiction, Young Adult and occasionally romance novels, but will consider any book. A unique feature of their reviews is "Old School Wednesdays," where they review books that have been published for at least five years. They also sell ads on their website. Because of their Klout score, a review by them would be influential.

Also, with blog and review sites, you may consider advertising with them in addition to requesting a review. The goal is finding a site that caters to your target market.

Beyond Klout

Besides Klout, there are numerous other ways to find influencers for your target market. In future chapters, we will discuss specific methods to use on various social media platforms, but for the already initiated, here are a few tricks to find influencers on the World Wide Web.

For Google+, influencers are found in the communities. Frequent posters and people with large followings are all potential champions for your books. The more your target market is connected to the tech community, the more likely G+ is the place for you to be.

Twitter is a great place to look for hashtags (#) connected to your genre—#SFBooks, #ChildrensBooks, #MysteryBooks, #Paranormal. As you scan the hashtags, which are metadata tags that denote topics or keywords, you may discover new people to follow. One of them could be your next influencer.

On Pinterest, you can search for words and pin categories. (We'll cover more about that in Chapter 8.) Find people who are pinning the latest covers or other items in your genre—and score! You may have someone that will be interested in your work.

So what is the next step after you find a few potential influencers? Spend some time getting to know them. Follow them on their social channels, check out their websites and engage with them and their community. When you decide it's time to contact them, make sure you do it appropriately. And by appropriately, we don't mean using correct grammar and being polite—though both of those are huge pluses. We are talking about determining how and where they like people to contact them.

For some social media users, it may have to be a direct message on Twitter. For others, they may have a website with contact information. Make sure you look closely because sometimes it is hidden at the bottom of the webpage. Can't find it? Check the "About" page, where they may give you an idea of how best to contact them.

Next up is your message—and in the words of Amy Schmittauer in her YouTube Magnet Minute video *How to Connect with Online Influencers*, "make it short and sweet."[122] Tell them who you are, what you want and why they might be interested. Or more importantly, why their audience would be interested and what you have to offer them in return. If they prefer e-books, don't offer to send them a print copy. If they love print books, spend the money and send them an autographed copy if you feel they will be a good influencer for you.

You should close with a "call to action." Tell them what the next step is going to be and don't leave them guessing. For instance, if they prefer electronic books, you can close your message by asking them to contact you with an e-mail address where they would like their book sent, and in what format they prefer. If they don't indicate a preference, just ask.

Let's work through an example. We'll use FollowerWonk. com, a free service that searches Twitter users' profiles for keywords. We did a search for "Paranormal Romance reader" and came up with a list of accounts that included Avon Books (a publisher of the genre), E. B. Black (an author) and Brandi (a reader, reviewer and blogger).

Brandi has over four thousand followers and has posted almost eighteen thousand tweets. Looking at her history, you see she is a fairly consistent tweeter and her bio tells you she is a reader and reviewer of Paranormal Romance, Urban Fantasy, Adult Fantasy and YA books. When you scan her tweets for content, you'll find she tweets mainly about reviews and book releases, but doesn't seem to be garnering much engagement from her followers, with few retweets and even fewer favorites. When you visit her website, it hasn't been updated since 2013, and neither has her Facebook page. Things start to look bleak—and when you go to her Goodreads account link from the website, you notice she has only reviewed one book this year.

We know what you are thinking: skip her and look for someone else, but let's not be hasty. Since she's such a very active tweeter, let's check what she's posting. After a little investigation, you realize the reviews she is tweeting about are not her own, but are from other bloggers. You see one for a paranormal book by an author similar to you, and when you click on it you discover a review by Author X on her website.

When you look at that author's website, you find she is a prolific reviewer of books with double-digit comments on her blog posts. She also adds her reviews to Goodreads and has share buttons that let you know her reviews are being shared by others on Facebook and Twitter. Her bio tells you she likes Paranormal Mysteries and romance and has recently discovered YA.

You pop over to her Twitter feed: sixty thousand posts and five thousand followers. Her Facebook page has almost three thousand followers and is updated regularly. When you search for her Twitter handle on Klout, you discover she has a score of 57, which is above average, and you can review some of her most successful posts.

That's how you fall down the rabbit hole in search of potential influencers. Much like scouring a bibliography at the back of a non-fiction book, you'll find new sources to investigate. Brandi didn't work out, but she led us to Author X, who may be a potential good influencer. Follow that author's Twitter feed, like her Facebook page and start to engage. See if she is a good fit and contact her for a review. If she declines, thank her for her time.

Keep in mind that while you're posting on the various social media platforms, the people you're interacting with—some of whom are on the other side of the globe—may be either influencers or brand advocates. Sometimes both. If you treat those people professionally and with courtesy, they'll spread the word about you and your books. And nothing is better than word of mouth.

Chapter Five Endnotes/Links

1. Smart Bitches, Trashy Books—(http://www.smartbitchestrashybooks.com/)
2. SF Book Reviews—(www.SFBook.com)
3. The Book Smugglers—(http://thebooksmugglers.com/)
4. Frenchy and the Punk—(http://www.frenchyandthepunk.com/)
   Abney Park—(http://www.abneypark.com/)
   The Extraordinary Contraptions—(http://www.reverbnation.com/theextraordinarycontraptions)
5. Steampunk World's Fair—(http://steampunkworldsfair.com/)
6. Nalini Singh—(http://www.nalinisingh.com/)
7. Sister Spooky—(http://www.sisterspooky.co.uk/)
8. Klout—(https://klout.com)
9. Klout Score—(https://klout.com/corp/score)
10. Top 5% of Influencers—(http://www.whatsupinteractive.com/blog/social/1442-your-klout-score)
11. The Book Smugglers Hugo Nomination—(http://www.thehugoawards.org/hugo-history/2014-hugo-awards/)
12. Amy Shmittauer—(http://youtu.be/Z_8mqrViHJI)

6 Facebook: Where Your Friends Are

When it comes to the social media kennel, Facebook is one of the big dogs. With over one billion users (five hundred million of whom use the site daily), it's an ideal platform where authors can build their brand. With forty-two million brand pages, it's also the most popular social media platform in the U.S. Seventeen percent of the time people spend on their personal computers is on Facebook. The demographics are spread pretty evenly across the sexes with the edge given to women. Ninety-eight percent of those users view content in their newsfeed, rather than bouncing from page to page. Not surprisingly, the fastest growing segment of those users are accessing Facebook via their mobile devices.

Facebook offers you two ways to connect with your readers: a **Timeline** (sometimes called a profile) versus a **Page**. The differences are reasonably straightforward.[1]

Timeline

A timeline/profile is for *non-commercial* use and must be registered under a personal name. Those who wish to follow your timeline will send a friend request, which may be accepted or denied. At present, you are limited to *five thousand* friends. After that, you will need to delete a few to add new friends.

You can also allow anyone to follow you and read all posts you make "Public." By going to the "Followers" tab in the Facebook settings, you can turn this feature on by selecting "Everybody." This way you can allow the "Public" to see certain posts while still maintaining a more private "Friends" list. There is also a "Follow Plugin" available for use on your website so people can easily follow your Facebook timeline.

While a timeline does not include all the cool analytical tools, you can customize it to show your favorite musicians, authors and books, television shows, movies and organizations. Your personal history (when you were born, where you went to school, relationship status) is available should you decide to share that information with those who follow your page.

Most authors use their timeline to post items of a personal nature, like their child's engagement or graduation photos, along with social commentary, memes and endless pictures of their pets. For others, however, their timeline is their only Facebook presence.

Under the Facebook settings general tab, you will find a line where you can customize the URL (web address) of your timeline. You can do this only once and the new URL should include your real name. For instance, Tyra's timeline is facebook.com/tyraburton, and if you go to facebook.com/janaoliver, you will reach Jana's author page.

Page

A page, on the other hand, is designed for businesses, organizations, celebrities or, in this case, an author. People "like" your page and there is no limit to how many Facebook users may follow you. For example, author Ilona Andrews has sixteen thousand followers while Anne Rice (author of the Vampire Chronicles) has over one million. Because both Andrews and Rice have substantial fan followings, they each have a blue "badge" that indicates Facebook has verified that their page represents who they claim to be. You cannot request to be verified, but if you reach enough fans Facebook may take notice and add the badge for you.[2]

Just like with personal timelines, you can customize the URL of a Facebook page. Log in to your account and go under the username section to make that change.[3] If you own more than one page, a dropdown menu will be there for you to choose which one you wish to name.

One of the biggest advantages a page offers is the "How are we doing?" data that reveals how many people are reading your posts, where they're from, and how much "reach" you are achieving.

A page is the perfect place to introduce and reinforce your author brand, as well as an opportunity to interact with your readers in a timely fashion. It is the ideal spot to keep your readers in the loop on future projects, appearances and your latest book news.

Which leads to the question, as an author, should you have a profile or a page? Some authors have both, restricting the business portion of their lives to their Facebook page while reserving the followers on their personal profile to close friends and family. Others readily share content from their author page to their private one. It all depends on how you wish to manage your online presence and if that data is important to you.

Several years ago, Facebook created an algorithm or formula they called "Edgerank." It determines if your post appeared in your friends' feeds, if it was a timeline post, or in your fans' feeds, if it was a page post. They have stopped using the name, but the formula still exists and has grown more complicated. We'll talk more about the formula later, but it needs a mention here because many people believe they are reaching fewer and fewer people with their Facebook page and more with their personal timeline. Unfortunately, we only have anecdotal evidence, because user statistics are not available for posts on personal timelines.

Another issue is how much of your personal life (or your career) should appear on either your Facebook page or timeline. This is a decision you have to make, and once you've decided on a strategy, it is best to keep that consistent. Again, be mindful that you are a public figure now, and as such, what you post on your timeline or page can go across this big world with one click, for good or ill.

> **Pro Tip: if you do engage folks using both a timeline and a page, ensure your author photos are different. That way your followers can easily differentiate the two accounts.**

Also, if you're posting on someone else's timeline or page, determine which of "you" is the poster. Facebook makes it easy to change identities by clicking the downward arrow at the top right of your page or timeline.

The Key Elements

If you choose a page (rather than a timeline/profile) to represent your brand to your readers, there are key elements that you need to consider when setting up that page. First and foremost, your presentation should be consistent with your brand. Some authors achieve this by mirroring the headers or images from their website, and even going so far as to use the same type fonts. That way you're telling your reader, "This is the place." Others work off a basic theme, which may be based upon their books' genre(s). That way there are similar elements between your page, your website and your other social media platforms.

First rule of the Facebook road, which pretty much extends to all the other social media highways: Don't just talk about your books. Readers want to know about you, how you write your story and do your research, as well as in-depth information about the characters. They do *not* want an infomercial. There are myriad ways to offer that kind of content without being the author who constantly shouts, "Buy my book!"

Creating engaging content is key for your page's success on Facebook. Only 6-15% of the fans of your page will actually see your post in their newsfeed. It is only through creating engaging posts or paying for advertisements or sponsored posts that you can increase your reach (the number of people who see your posts). You need to understand how Facebook chooses what goes on each individual's newsfeed.

Time matters. The older your post, the less likely it is to be placed in newsfeeds. If a person has interacted or engaged with your posts in the past by liking, commenting or sharing them, then that helps increase the probability they will see future posts.

If the post is popular it will have an increased likelihood of showing up in others' newsfeeds. In fact, if you have past posts that are highly popular, that increases your chances of getting into people's feeds. Also important is if the post is the same type that has proven popular in the past.

Facebook's goal is to fill timelines with high-quality content from people and pages that matter to them. It's all about creating content that is engaging and encourages interaction. We also know that the last fifty people or pages a person interacts with on Facebook are more likely to show up in their newsfeed. What does this mean for you, the author? You probably want to consider posting at least one to three times a day, depending on when your primary target market is online, in order to be sure your content is recent and fresh. You also want to consider utilizing photos whenever possible, and give people a reason to like, share and comment.

Some of the ways you can achieve maximum reader engagement:

1) Use images/videos that resonate with your book's theme

An example: you write Scottish historicals. Bless you! So why not include a video of a brawny lad showing us exactly how to put on a traditional kilt (which is one lo-o-ong piece of fabric, minus the sewn pleats, and not as easy to don as you might think).

Or if your hero/heroine is Irish, on St. Patrick's Day, post a video of Border Collies herding a bunch of lads through an obstacle course. It's funny and ties right back to the holiday and maybe your character or book.[4]

> **Pro Tip:** Use vertical images as they take up more space on your followers' newsfeeds.

2) Create Hashtags

Add a hashtag (#) to create a clickable link to other people's posts that contain the same subject. #Scotland, for example, brings up posts related to that topic including links to articles, cartoons and gorgeous pictures. You can do the same for your books' themes and locations, which helps spread your posts beyond just your usual readership.[5]

3) Create your own quizzes or puzzles based on your characters and/or books[6][7]

4) Create your own YouTube videos and share the links

These videos may be you talking about your fave characters, about the books or movies you love or about your hobbies. You can also create videos of some of the locations used in your books, like P.C. and Kristin Cast's tour of Tulsa for their House of Night series.[8]

5) Share a Google map of important book-related locations[9]

6) Post snippets from your books, especially leading up to a new release

7) Offer a contest (after you've reviewed Facebook's Terms of Service and their Guidelines page, which is located under promotions)[10]

Readers adore giveaways! Facebook has relaxed its requirements for contests, but you may not use personal timelines to administer promotions or contests, or ask your fans to share on their timeline or a friend's timeline to be entered. If you plan on running promotions, you will want to choose a page over a timeline for connecting with your fans.

8) Share photos and comments about your research trip for your latest book

9) Post items of personal interest

Whether it is your appreciation of genuine Texas barbecue or your My Pretty Pony collection. Readers *love* this kind of interaction. One of the co-authors of this book collects amusing cartoons, quotes and pictures to use when a topic she wishes to post matches one of those images.

10) Share your blog posts on Facebook

By doing so, you will drive readers to your website where they can find more information about you and your books. Do make sure you to include a horizontal photo in your blog posts. It will better highlight your blog post on people's newsfeeds.

In essence, you are creating additional content beyond your books to bring readers further into your fictional "world." Successful authors are all about giving readers a great experience and stoking interest in their next book. In this case, Facebook's massive reach works in your favor. In addition, such posts can drive traffic to your website or blog, which is always a good thing.

Besides creating your own content, Facebook readily lends itself to sharing other people's posts, which is exactly what they want you to do. Employ this share-ability to your advantage. See a post or an image that allows you to mention an upcoming appearance in a different state or country? Use it. Share posts from your fellow authors, be it about their writing journey or their own books, because it's not about competition, but about sharing a pool of eager readers.

Facebook really wants you to share images and video, and because of that, you don't receive as much "visibility" if your post doesn't contain one or the other.

> **Pro Tip:** Once you've placed the URL in the post and Facebook has located the page, strip the link out to keep the post from looking cluttered. The link will remain in place.

Unlike on a timeline, Facebook pages allow you to schedule future posts, which is a handy feature when you're traveling to an event, flying across the country or head down in revisions. You can stay in touch with your readers even if you're nowhere near a computer. Just remember to go back and like and comment on their responses. Because what's the point of posting on your favorite author's page if you are only going to be ignored?

Organic vs. Paid Reach

According to Facebook's web page: "Organic reach is the total number of unique people who were shown your post through unpaid distribution. Paid reach is the total number of unique people who were shown your post as a result of ads."[11]

Facebook clearly has a strong platform, and one that continues to grow daily. However, reaching your audience isn't quite as easy as it once was. Remember, after recent changes to Facebook's super-secret algorithm, only 6-15% of your total followers see your posts.[12]

Part of this is because of the sheer volume of posts, which would easily overwhelm your readers if they were all listed in their feeds. Facebook's primary goal is to create a high-quality experience for their users. The other reason is that Facebook is a publicly traded company that needs to turn a profit, and the best way to make their investors happy is by encouraging you, the page holder, to "sponsor" your posts in order to ensure they are seen by your target audience. That's why Facebook constantly improves ad layout, the size of ads in the right hand column and so forth.

To see what kind of organic reach your posts are receiving, check out the number in the left-hand bottom corner of each of the posts. We'll break those numbers down further in a moment.

While organic reach doesn't cost you a dime, sponsoring your posts does. But is it worth the extra expense? We believe it is best to choose carefully which posts you feel need an extra boost, which may include upcoming appearances, book tours, new books and really important news. Again, it all depends on your ad budget.

To promote or boost your timeline post, click Promote, which is next to the Like, Comment and Share links. Unlike a page post, you cannot customize who will see this post. On your page, click on the blue Boost Post rectangle at the bottom right of the post. At that point you will be presented with two options in terms of audience: people who like your page and their friends, or people you choose through specific traits and geography.

The first is fairly straightforward: your friends and their buddies will see your sponsored post in their newsfeeds.

The second option is more intriguing: you can choose who you wish to see your post, indicating a choice of geography (country, state/province or city), an age range and gender. You can also generate a list of interests, such as Romance Novels or Suspense Novels. This will allow you to target your post to those who meet your specific target audience.

Then you are prompted to choose how much you wish to spend, and with that number in mind, Facebook will tell you approximately how many people will see your post. Under the More Options section, you can decide how many days (up to seven) you want this post to be promoted. Set it and forget it (once Facebook has either approved or declined your request).

You can also promote a post on a personal timeline so

it will appear higher in your friend's newsfeed, but will be marked as a sponsored post. As mentioned previously, pick which posts to promote based on your target audience and timeliness of the post. Then give it a try!

Which leads us to Facebook analytics, or…more of that math stuff. Trust us when we say that this kind of math is *very* important.

Facebook Analytics

One of the most difficult things for an author to determine is which advertising/promotional methods work vs. those that fall flat. Wasted time equals less time for writing, so data is your friend.

Facebook's page analytics highlight which posts your readers are enjoying/sharing, and where those readers are located. It's in an easy-to-digest format, with no mathematics degree required.

It's also private, meaning only you as the page administrator can see the data. If you have your own Facebook page, follow along as we check out the different sections.

At the top of your page is a link entitled "Insights." This is where you enter the world of magical numbers. Under that link is another series of choices, including "Overview," "Likes," 'Reach," "Visits," "Posts" and finally "People." Under these is stored information about your page's performance, including Page Likes, Post Reach and Engagement. Inside each of those boxes is an arrow that takes you to further interesting data, but we'll get to those in a second.

Beneath those boxes is a section that shows your last five posts (or more if you click the link below). Here you can view your reach and engagement plus the option to sponsor (or promote) any of those posts. Also, underneath that section is an option to watch other people's pages to contrast their performance with your own. Inside each of these boxes is a benchmark option located to the right of the screen. Be sure to check it out.

So how does this work in the real world? Let's use an example. On Jana's page for the week of July 1-7, 2014, the numbers tell us that she had a total of 3500 page likes up a bit from the week before, and that she scored ten new likes in those seven days, three on the first of the month and five of them on July 2nd.

Why is this important? Because knowing this allows her to go back to those dates and check which posts gathered that sort of attention and a new "like," the ultimate approval. Also, she may be able to tie this increase to a touchpoint, like a guest blog or a book signing.

On July 1st, she posted a map that pinpointed many of the Atlanta locations mentioned in her Demon Trappers series, and on the 2nd, she posted a link to Page to Premiere, a site that creates wish lists of books they hope will make it to the big screen. Since that blog post included pictures of Jana's top picks for actors to play the characters in her series, this was very popular with her readers.[13]

Page Likes

A click on the arrow in the Page Likes section shows you how many likes you acquired today, then below is a graphic indicating "net" likes (those added or removed) and where those likes came from (on your page, on your posts or via a mobile device).

Reach

The next column over breaks out the reach per post. In total, Jana had a little over twelve thousand total reach, up over 1000% from the week before. Below that a graph breaks out the post reach, showing the difference between this week and the week before.

A click on the arrow takes us to the section that shows us organic vs. paid reach (again per day), likes, comments, shares and then unlikes and total reach. On July 1st, Jana reached 6800 folks according to Facebook's stats.

Engagement

The final box in this section is Engagement, which, according to Facebook, is "The unique number of people who liked, commented, shared or clicked on your posts." Those numbers are broken out into those four categories and if you use the arrow to go to the data you will find it mirrors that found under the "reach" section.

Visits

The Visits section shows you the number of times your page tabs were viewed, and there is also a graph for Other Page Activity, including how many people have commented or posted directly on your page. And finally, there is External Referrers, which shows you the source of people visiting your page. In Jana's case, it was mostly Google.

Posts

This is one of the most important analytic sections. You can spend a good amount of time pondering the posts' data—but here are the things you want to pay particularly close attention to.

1) Look at your page's activity by days of the week. Are they all about the same? Do you have better/worse engagement on one day versus another? Knowing which days your posts are most successful helps you know when to time big announcements in the future.

2) This is where Facebook drills down to hourly data, which is a useful tool that shows you when your fans are most likely to read and engage with your posts. If you have an international following, these numbers may show peak times for Europe, Australia and other parts of the world.

In Jana's case, her peak engagement time is between noon and five p.m. in the Eastern U.S. time zone, which means her main engagement is most likely with readers in the U.S. and Europe.

To verify if this is correct, she only need check out the People section, which we will discuss down the line.

3) Next is the "All Posts Published" section, and it contains a wealth of information. Each post is listed with the date, time and type of post. With the columns following these, you can start to determine which type of post and time of day works best to increase reach and engagement.

4) The reach column displays the total number in two different formats—organic/paid or fans/non-fans. If you are doing a paid campaign, you can compare the reach you are getting organically with what your campaign is driving. With the fans/non-fans, you are able to see how your reach is extending beyond the people who like your page.

For instance, on the Facebook page for our local writers' group that has 575 likes, we had a post with a total reach of

833. Our fans comprised ninety-one of the total reach while non-fans were a staggering 742.

How did this happen? Let's look at the next column to find out.

5) The engagement column tells you how much and what type of engagement each post is garnering on Facebook. There are four different ways to look at this data from the dropdown menu. First is the "Post Clicks/Likes," "Comments and Shares" section. This is especially good to look at if there was a link involved with the post, as it will tell you how many people clicked on it, as well as the total number of likes, comments and shares.

The second menu is "Likes," "Comments" and "Shares" where the analytics breaks each of these out into separate numbers. Going back to our post with a reach of 833, we find that fifty-five people liked it, nine made comments, and six people shared it. When someone engages with your post by liking or commenting on it, it has the potential to show up in their friend's newsfeed as well as the activity "ticker" in the upper right-hand corner of the home page. When someone shares a post, all their friends have the potential to see it. This is how you can increase engagement by providing your fans the type of posts they like, at the right time, and on the right day.

You want the next set of numbers in the dropdown menu to be low, because it focuses on how many post hides, hides of all posts, reports of spam and unlikes of your page happened corresponding to each post. This can help you determine what your fans find annoying or dislike. Particularly you want to pay attention to posts that are marked as spam or generated several unlikes of your page.

Lastly, you can choose "Engagement Rate" from the dropdown menu to get a percentage of the people who were reached by the post that also engaged with it in some fashion. For a post with a reach of 833, 8% engaged with it. We had another post with only a reach of sixty-seven but 10% of those that saw it engaged with it.

As you can see, there is a lot of detail you discover in the post analytics section and with just a few minutes spent each week, you can determine when and what to post for greater success.

People

This section is most helpful because it breaks out your follower demographics according to sex, age and where they reside. As with some of the other sections, this area has subsections ("Your Fans," "People Reached," "People Engaged") and the numbers vary for each of those.

In Jana's case, under Your Fans, we find that women are 93% of her followers (as compared to "all of Facebook" at 46%). For men, it's 6% (versus Facebook's 54%). Which tells us that Jana's draw among females is *twice* that of Facebook, while males don't follow her page as often as Facebook's standard. Curiously, the total number of men and women who follow Jana doesn't quite add up to 100%.

What about age? Jana is best known for her Young Adult books, but only 17% of her followers are between 13 and 17 years old. Part of this is because teens prefer other social media platforms to Facebook. For her, the 18-24 range is the strongest at 31%, and 25-34 is 24%.

Which means that over half her followers are in the 18-34 age range. That is important to know as it indicates what types of subject matter might be of more interest to those folks.

Under those other two tabs (People Reached and People Engaged), you will find how those age numbers hold up in regard to what you're posting. In the case of People Engaged, curiously, Jana's male followers were more engaged (19%) than their small numbers would suggest, which means those men who do follow her page are much more likely to comment on or share her posts.

Finally, the geographic breakdown provides a snapshot of your "fandom" worldwide. In Jana's case, her strongest two countries are the United States and the United Kingdom, with Australia and Germany following right behind. Given these two of these countries include her bestselling markets, it's not much of a surprise that her Facebook followers reflect that.

Even better, the geographical listing breaks out your followers by *city*, both domestically and internationally, and by *language*. Knowing where your fans live and what language they speak is very helpful so you can use content that is relevant to them.

Post an article of interest to your French readers on Bastille Day, or to your Welsh readers on St. David's Day? Or how about a post about the Iowa State Fair? That kind of attention to your readers shows you truly care about them, no matter where they live in this big world.

Why Facebook Matters

Though this big social media dog can be frustrating at times, its greatest strength is its ability to directly connect you with your readers, with the ability to accurately tailor your posts to match those readers' expectations.

In the end, a savvy Facebook strategy will nurture your relationship with your fans, gain you new readers and help build enthusiasm for your next book launch.

Chapter Six Endnotes/Links

1. Timeline vs. Page—
(https://www.facebook.com/help/217671661585622)
2. Facebook Page Verification—
(https://www.facebook.com/help/269614913183739/)
3. Facebook Page URL Customization—
(https://www.facebook.com/username)
4. St. Patrick's Day Video—
(http://twentytwowords.com/sheepdog-herds-men-to-the-pub-in-a-funny-guinness-commercial/)
5. Facebook Hashtags—
(https://www.facebook.com/help/587836257914341)
6. Survey Monkey—
(https://www.surveymonkey.com/mp/tour/gettingstarted/)
7. Puzzle Maker—
(http://www.puzzle-maker.com/crossword_Entry.cgi)
8. House of Night Video—
(http://www.houseofnightseries.com/tour-of-tulsa/)
9. Demon Trappers Location Map—
(http://demontrappers.co.uk/demon-trappers-atlanta-map/)
10. Facebook Terms of Service—
(https://www.facebook.com/page_guidelines.php)
11. Organic vs. Paid Reach—
(https://www.facebook.com/help/285625061456389)
12. Social Media Today—
(http://socialmediatoday.com/tara-urso/2281751/your-facebook-pages-organic-reach-about-plummet)
13. Page to Premiere –
(http://www.pagetopremiere.com/2014/07/page-to-premiere-wish-list-the-demon-trappers-by-jana-oliver)

7 What's All This Tweeting About?

> Therefore, *since brevity is the soul of wit,*
> And tediousness the limbs and outward flourishes,
> I will be brief. Your noble son is mad. . .
> *Hamlet Act 2, Scene 2, 86–92*

If you're delivering bad news, brevity is important. After all, the Bard was able to do it in just 135 characters. That same brevity applies to Twitter, a powerhouse social media site with nearly a billion registered users and 255 million people who regularly use the site each month.

Sixteen percent of U.S. adults employ Twitter for their online interaction, dividing right down the middle between male and female. This makes Twitter the perfect social media "home" for nearly everyone, ranging from celebrities, authors and everyday folks going about their daily lives.[1]

Twitter is a platform all about the *here and now*. It's real-time interaction. You can never "catch-up" on Twitter's feed, at the very best you can drop by and see what is going on right this minute. Tweets can be as simple as announcing the train is late *again,* to news of a countrywide revolution.

When the Bronx Zoo's cobra went on walkabout a few years back, some enterprising person immediately created a Twitter "handle" detailing the snake's hilarious exploits in the Big Apple. That account currently has over one hundred and seventy-four *thousand* followers.[2]

No surprise, the Bard has found a home on Twitter, sharing his wit and wisdom. Or in this case, celebrating his 450th birthday.

William Shakespeare @ShakespeareSays - Apr 23
"Prepare for mirth, for mirth becomes a feast: You are princes and my guests." #Shakespeare450 #HappyBirthdayToMe

And not to be outdone, Southwest Airlines chimed in…

Southwest Airlines @SouthwestAir - Apr 23
To fly, or not to fly: there is no question. Fly… Duh. #HappyBirthdayShakespeare

But you don't have to be a company or a famous playwright to tap out your message, such as this anonymous tweet making the rounds:

Where would I be without my mother? Probably in the middle of traffic, without my jacket on, talking to some stranger.

Or this one…

Jana Oliver @crazyauthorgirl - Jul 3
Getting my Twitter on in the social media book I'm co-authoring w/ @tyraanneburton #AmWriting

The Purpose of Twitter

Twitter's beauty is that it's limited to just 140 characters (counting the spaces as well). That's not much wordage, so you must learn how to choose each word for maximum effect. For Jana, who came from a copywriting background, this comes naturally. For others, it's much harder. And for some, Twitter just doesn't work at all. As with all social media, pick and choose what you're best at and politely ignore the rest.

Twitter is most effective when you are in the moment. Though fun to use on a personal basis, Twitter offers genuine benefits for an author, because it's a quick way to send out news and stay in touch with your readers. If your readers are international, Twitter can be a great choice because 77% of their active users are outside the U.S. Unlike lengthy blog entries, Twitter doesn't chew up a lot of time.

To see someone's tweet in your feed, you need to follow them. Unless you set up your Twitter profile to be private (meaning you have to allow people to follow you and see your tweets), what you post on Twitter can be seen by anyone.

If you go to www.twitter.com/tyraanneburton, you can read the tweets Tyra has made on Twitter since she joined (if you have enough time). As with all things on the Internet, what you post is there forever.

> **Pro Tip: If you really do need to post a longer post, you can use one of the programs such as Twitlonger, which will allow you to run over the 140-character limit.[3] Or, better yet, write your longer message as a blog post and link to it in your tweet.**

Logistics

Like any social media platform, there are rules of the road regarding Twitter. Let's start with the basics—the @ symbol, RT (retweet), DM (direct message) and the hashtag (#).

The @ symbol is used before someone's "handle" in order to tag them, or in Twitter lingo, "mention" them. If someone uses @crazyauthorgirl in a post, you can click on it and it will take you to crazyauthorgirl's Twitter page.

The @ can also change who will see your post.

Jana Oliver @crazyauthorgirl
@tyraanneburton Our book hit the New York Times @nytimes bestseller list! #FTW

Okay, that message is totally fake, but still, who would get to see it? People who follow *both* Jana and Tyra. Everyone else on Twitter is excluded from having it in their feed.

How do you get the message to those folks who don't follow both of us? By putting some sort of *character* in front of the @ symbol. The most common used character is a period, though it can be a word as well.

Jana Oliver @crazyauthorgirl
.@tyraanneburton Our book hit the New York Times @nytimes bestseller list! #FTW

Or…

Jana Oliver @crazyauthorgirl
WOOT! @tyraanneburton Our book hit the New York Times @nytimes bestseller list! #FTW

Now more people see this tweet and we might just be able to start a bit of word-of-mouth action. If someone wants to forward it to their followers, they hit the retweet symbol (which looks like a square composed of arrows).

Chris Jasper @CJTrapper
RT.@tyraanneburton Our book hit the New York Times @nytimes bestseller list! #FTW

At this point the period is no longer needed and can be removed, as the RT serves as characters before the original tweet. And just to further confuse the issue, depending on what program you use, the RT may not show in the actual message, but in a line above the tweet. Either way, the effect is the same.

If your tweet is one you are hoping gets RTed, consider making it 120 characters or less. This allows room for the RT and for a comment by the person reposting it.

But what if Jana only wants Tyra to see this message? Then Jana will send her a DM by putting a lowercase d in front of her username, but without the @ symbol.

Jana Oliver @crazyauthorgirl
d tyraanneburton Our book hit the New York Times @ nytimes bestseller list! #FTW

This way *only* tyraanneburton sees the message. One word of caution: If you are DMing someone back and forth, be sure to include the d or others will see your message. If it's of a highly personal or controversial nature, that can be embarrassing. You can also use the DM function (icon in the top right corner if you are at twitter.com) to do this for you and minimize the chances of spreading news you wanted to keep private.

The hashtag (yes, it's really a pound sign) is a way to group tweet under a certain topic or keyword. When someone searches for tweets containing a certain hashtag, all the posts containing it will be included in the results. In the above example #FTW (for the win) is a popular hashtag under which you'll find comments about awesome sports figures, gaming, good deeds and other uplifting news. During the World Cup finals, the various matchups were given their own hashtag, such as #BrazilvsGermany.

The Romance Writers of America annual conference generates an energetic stream of tweets. Attendees follow the latest news by setting up a separate thread: #RWA**. (The asterisks are replaced by the last two digits of the current year.) For example, #RWA14 had a steady stream of comments for months in advance of the conference, then during and after the event. It's like a Twitter "neighborhood."

As an author, you can set up hashtags of your own, as they are not exclusive (which means the one you use may mean something completely different to someone else). Jana set up a #DemonTrappers hashtag which allows her readers to follow any tweets that fall under that category. Our local romance writers' group uses #GRW for tweets about our meeting, but there is a German company that uses it, and it was also used during the World Cup. There is no place to "reserve" a hashtag. People can use whatever hashtag they want and sometimes that makes things confusing.

During her *12 Days of Daemon Black* promotion, author Jennifer L. Armentrout (and her publisher Entangled Teen) used the hashtag #daemoninvasion to direct her readers to the latest photos and news. This helped her readers easily find content that may have gotten overlooked in the real time Twitter feed.

Getting Started

Signing up for a Twitter account is pretty painless though choosing your username can be tricky, as it has to be unique —Twitter can't handle one hundred Stephen Kings. Do keep in mind that when other users retweet your messages, your username is included in that 140 word count, so if you have an *extremely* long username, that's fewer words available for the message. Authors can either use their own names or some variation (like Neil Gaiman, who tweets as @neilhimself).

Once you've set up your account, be sure to fill in the profile information, including a short blurb about who you are—and make it intriguing so more people will be interested in following you. Also, make sure to include your website. To get ideas of how to do this, check out other folks' profiles. Then upload an image for your avatar and one for your Twitter page cover. If possible, tie these back to the images/look you've chosen for your website and your blog.

Oreo does a great job of branding all of its social media to match its website.[4] In the publishing world, author D. B. Jackson does a good job of branding his historical fantasies across multiple platforms utilizing the primary images on his website.[5]

Often authors will change the avatar to reflect their current book, which is another quick bit of marketing. Just note, when you change the avatar it changes over all your tweets, even those in the past.

Twitter Uses

There are various situations where Twitter can come in handy, and not only for having short conversations with your friends and readers. Jana often tweets events, not only before that event begins, but during and after. If she's on a panel during a convention, she'll tweet the time, place and details. Same for a signing or other personal appearances. Often those tweets help her readers drop by for a chat.

She also tweets her various trips, offering little snippets about travel, observations about the countries she visits and the people she meets. When Jana is on the road, she holds

Meet-and-Greets at a specific location and time so she can chat with her readers outside the confines of a convention. It often happens, despite all her research, that the location she picked for the meet-up is not an option, despite whatever that location's website claims for hours and days of operation.

In the case of a meet-up planned in Brighton, England, the site had been closed for a couple months, though the website was still live. Twitter to the rescue! Jana announced the news, quickly picked another location and routed her readers to the new address.

Another great use for Twitter is to have your characters tweet with your readers (and between each other). This requires you to set up individual accounts for each of the characters, which can be a bit of a pain, but readers really enjoy this kind of interaction. Author Jeri Smith-Ready (The Shade Trilogy) had her two heroes tweet back and forth in a rowdy give-and-take. Fans of Team Zackary (he's Scottish) and Team Logan (he's American) quickly united behind their favorite character.[6] Jeri is a serious tweeter (with nearly forty-six thousand posts to her name) who often offers play-by-play commentary on football games and other sporting matches. Her fans adore her for it.

Twitter has a few tricks up its sleeve: once you've posted, select the tweet then click the three dots at the bottom of that message. This gives you three options—share by e-mail, embed the tweet (by providing a code you can place on your website, for instance) or "pin" it to the top of your profile page. In particular, the last option allows you to keep vital information at the top of your page where folks can find it. You are only allowed one pin at a time, and when you choose a new one, the old one is automatically replaced.

One question we are frequently asked is if you should forward your tweets to Facebook so they will show up on your timeline or page. We're not fans of that type of sharing, because we believe that each social media platform requires its own kind of content. The other downside of automatically posting your tweet is that it tends to clutter up your timeline/page if you're a prolific tweeter. Also, the wording used on Facebook and Twitter is different. Even Martha Stewart uses shorthand on Twitter to fit a message into 140 characters.

On Facebook, you aren't limited in length, and photos are much more important. So, you may post about your latest guest blogging on both platforms, but on Facebook you can do a bigger lead-in and include a link with a photo.

On Twitter, you'll do a short lead about it and include the link, hopefully shortened using a service like bit.ly.[7] All those quizzes and memes that are all the rage on Facebook? They barely make a blip on Twitter. Two different platforms, two different styles of content.

Twitter Housekeeping

It's great to follow new folks—and trust us, it's addictive—but eventually your list will become unwieldy. Which means that when those folks do tweet something important, it's very likely you'll miss it because of the sheer volume of tweets in your home feed. Various applications have been created to help you tame that unruly list, and Mashable has them all in one nifty post.[8]

Tweeter Karma checks through your followers list and delivers a report as to who is following you back, and those folks who are following you but you're not reciprocating the love. If you have a ton of followers, this app can be slow.[9]

Now is a great time to mention our favorite Twitter client, Hootsuite, an application that really makes Twitter easier to manage.[10]

Hootsuite handles both personal and business accounts, and there are paid and free options as well. It has a dashboard under which you can set up your various Twitter feeds. The steams can include a feed for tweets that mention you (where someone has used @yourtwitterhandle in a post), one for direct messages, a "home" feed that will show all the posts by people you follow and finally, a "sent" feed for all the tweets you've posted (except for your direct messages).

You can also add customized feeds—Jana set up one to track her British publisher's Twitter posts, @MyKindaBook, and the posts of various friends. She also has one to track her #DemonTrappers hashtag in case someone doesn't address the tweet to her. During the previously mentioned Romance Writers Conference, we both set up feeds that tracked the #RWA14 stream, to stay on top of any special news. Though, to be honest, trying to follow hundreds of tweets is a bit daunting.

After you've set up your feeds, you can move them around the dashboard into new positions, which is handy if one feed has more importance on any given day.

Hootsuite has a few other nifty functions. Once you've created a tweet, check out the row of icons below it. There you can include an image or a file or add a URL, schedule a tweet, add a geographic location, choose which country's followers you want to see this message and, if you wish, a specific social media platform on which to clone that tweet once it's posted.

The scheduling feature is a really useful tool when you're busy, like at a book signing, a convention, on the road or buried in deadlines. It allows you to tweet at specific times, which is important depending on whom you're targeting that tweet towards. You will find there are peak times of engagement for your tweets, and you should work hard to offer comments during those times. The scheduling feature makes that process easier.

As for the geotagging options—this is a common feature on a number of social media platforms and should be used with caution. Not everyone is comfortable having the world know where you are when you're posting. If you're one of those who feel it's best not to tag your tweets (and other social media interactions), then turn that tagging option off.

Hootsuite isn't just for Twitter. You can manage your Facebook personal timeline and page using the service, as well as LinkedIn, Instagram and your Google+ business page (but not your personal one). It is available in desktop, iOS and Android platforms, and will sync across those devices.

Advertising

Much like Facebook, Twitter offers its users the option to promote either their account or specific tweets. The promoted account is suggested to other Twitter users who aren't currently following them in the hope that those people will follow that account.[11]

A promoted tweet is one where you want that message to reach a greater number of people than it usually would. You can also target a specific group of people. Promoted tweets are best coupled with a 'call to action' that urges the reader to do something to engage with that message. For example, if you're running a special contest to give away copies of your latest book, a promoted tweet would tell your readers that news and indicate where they need to go (your website, perhaps) to register for that giveaway. Or if you are going on a book tour, then a series of promoted messages highlighting the towns where you'll be appearing might be a good way to get the word out to your readers.

As with all advertising, you do need to determine if the dollars spent have helped you achieve the goals you set before you began the promotion. You did set goals, right?

And...just in time to help you with that assessment is Twitter's announcement that they're now offering that kind of information for their business accounts as well as basic stats on individual accounts.

Show Me Those Numbers

As we were going to press, Twitter announced that it would be offering analytics so that businesses (which you are) can more effectively judge how their tweets are performing.

According to Twitter product manager Buster Benson, "For the first time, advertisers will be able to see how many times users have viewed and engaged with organic tweets, so that they can more effectively optimize their content strategy."[12]

The kind of information you'll have access to includes how many times a tweet is viewed, how that tweet is performing in real time, the number of retweets, follows, link clicks, replies and how many times that message has been deemed a "Favorite." You can also download this data into a spreadsheet.

Since this is such a new development, please check our website under the Twitter section to learn more about the analytics as it rolls out. Be sure to follow us at @socialmuses.

Since numbers can be your friend, we recommend you try a free service called Twitonomy that will tell you a bit about any public Twitter profile.[13] If you look up your own, you will find a good snapshot of your Twitter history, including what days and times you post, how many RTs and mentions you receive, as well as what hashtags you most frequently use. Twitonomy has a lot of data—if you have the time.

So the final word on Twitter in less than 140 characters?

Social Media Muses @socialmuses
Twitter keeps savvy authors in touch with their readers, builds author brand & helps sell more books #FTW

Chapter Seven Endnotes/Links

1. Twitter Demographics—http://expandedramblings.com/index.php/march-2013-by-the-numbers-a-few-amazing-twitter-stats/#.U8U2NpRdXng
2. Bronx Zoo's Cobra—https://twitter.com/BronxZoosCobra
3. Twitlonger—(http://www.twitlonger.com/)
4. Oreo—(http://www.oreo.com/default.aspx)
5. D.B. Jackson—(http://www.dbjackson-author.com/)
6. Jeri Smith-Ready—(http://www.jerismithready.com/)
7. Bitly—(https://bitly.com/)
8. Twitter Management Options—(http://mashable.com/2009/06/09/organize-twitter/)
9. Tweeter Karma—(http://dossy.org/twitter/karma/)
10. Hootsuite—(www.hootsuite.com)
11. Promoted Accounts—(https://support.twitter.com/groups/58-advertising/topics/261-promoted-accounts/articles/282154-what-are-promoted-accounts)
12. Organic Tweet Analysis—(https://blog.twitter.com/2014/introducing-organic-tweet-analytics)
13. Twitonomy—(http://www.twitonomy.com/index.php)

8 Pinterest: The New Wishbook

At its simplest, Pinterest can be thought of as a visual bookmarking system, and as a planning/shopping board at its most complicated. It is often likened to a modern-day scrapbook where instead of cutting and gluing pictures and articles into a notebook, you are curating (or collecting) content online.

In general, users (pinners) "**Pin**" photos of a product, recipe or article, or cute photos that are linked back to a website. Pins can also be created from user uploaded photos. A user's pins are grouped together on **Boards** that contain similar content. Users can choose to follow all the boards of a fellow pinner or just certain ones. Unlike in some other social media, the "conversation" on Pinterest is all about the visuals.

A Pinterest feed will be populated by content from the people you follow, promoted or pay-per-click pins, and as your own content. People will like others' pins and even "**Repin**" them if they want to keep track of a pin on their own board. Top browsed categories include Food & Drink, Do-It-Yourself/Crafts, Home Décor, and Holidays and Events. Film, Music and Books is ranked seventeenth in the favorite browsed categories.[1]

With fifty-five million active *monthly* users, Pinterest has shown staggering growth over the last few years and now claims over one-third of U.S. women are pinners.[2]

> **Pro Tip: Be sure to visit our website for an easy visual explanation of Pinterest.**

First, let's walk through an example to help explain how Pinterest actually works.

You've signed up, have created a few boards, including one titled "Food," and you have started to pin as well as following other pinners. In this case, you make sure to follow your friend Sally's boards, because she convinced you to join the fun. One day, while browsing your Pinterest feed, you notice she has pinned a recipe for homemade organic peanut butter cups. The photo makes you drool, so you click the "like" button so Sally knows you found it interesting, then you click on the photo itself and it takes you to the site where you can check out the recipe.

The recipe is from a blog post at the Organic Treats site and the article walks you through each step on how to make them. After looking at the recipe, you decide you want to keep track of it but you don't want to refer to Sally's board each time, so you "repin" Sally's photo of the peanut butter cups onto your "Food" board. In the description box you type, "must try." When you are ready to make the peanut butter cups, you go to that board, and click on the photo, which will take you to the Organic Treats blog post (unless they have removed it), where you will find the recipe.

That's Pinterest in a nutshell.

Heed our warning—it is addictive. Hours can be lost as you explore boards and check out pins. People plan their meals, exercise plans and weddings on Pinterest boards. They find inspiration in the crafts and DIY projects while discovering new cooking methods. Don Faul, head of operations for Pinterest, says the platform is all about discovery—"Finding what you did not know you were looking for."[3] With a total of thirty *billion* pins, there's a lot to find.

As an author, why should you be excited about Pinterest? Isn't it just another time waster? If your books are primarily targeted toward women and if that female target market is between 15 and 39, you should run, not walk, to your nearest computer to start your pinning career. Why?[4]

- 86% of Pinterest users are female
- 63% of their active daily users are 15-39
- 39% use Pinterest instead of search engines
- 54% of the destinations of daily click-throughs are to blogs.

In Pew Research's *A Snapshot of Reading in America in 2013* study, women were statistically found to read more books than men, and a higher percentage of all women have read a book during the last twelve months.[5]

During a 2013 Random House publishing event, Carl Kulo of Bowker Marketing Research stated that women account for 60% of paperback book sales and 65% of e-book sales.[6] With women being the driving force in the book market and Pinterest's predominately female users, the two simply go hand in hand. One of the best things about Pinterest is that creating pins, repinning and liking are much more common than commenting, so less time is involved.

Tyra's most popular pin is two years old and still has engagement on a *weekly* basis. As of June 2014, it has 851 repins and 178 likes with—drumroll please!—three comments. That's it. *Three*. This means you don't have to respond at nearly the same rate as on other social media platforms like Facebook or Twitter. This means more time for content and less time for maintenance and, hopefully, more time for writing—if you set a clock to remind you to get off Pinterest. Remember, we told you it was addictive.

Using Pinterest

Pinterest has individual and brand pages, and you can sign up for both. There are certain benefits to having a business account including access to their analytics as well as the ability to do pay-per-click promotions. In order to reap the benefits of a business account, you must be verified. Pinterest requires you to verify either with an **HTML** (HyperText Markup Language) file or a meta tag (www.business.pinterest.com).

After you have been approved, you will see a checkmark next to your name and you will be able to access your Pinterest web analytics, as well as pay-per-click options when promoted pins open up to all business accounts. You can sign up for either or both types of accounts at pinterest.com. We've included "visual" instructions on how to do this on our website.

The first place you should visit is your "Account Settings" (www.pinterest.com/settings while logged in.) There you are able to fill in or change several sections including "About You," "Location" and "Website."

You'll also be able to set e-mail notification preferences, whether to connect and post your activity to your Facebook timeline, and whether to connect with your Twitter, Google+, Gmail and Yahoo! accounts. We suggest turning off "post activity to Facebook," as it can clog your account if you post a lot at one time.

After you have opened your account and fiddled with your settings, the next step is setting up a few boards. Remember, just like with other platforms, you shouldn't be all about marketing your brand or books on Pinterest, and your board choices should reflect that. Below are our suggestions for a good collection of starter boards. If you are already submersed in to the Pinterest world, make sure you have these key boards in your social media arsenal.

Popular Categories

The four most popular categories on Pinterest are: Food/Drink, DIY/Crafts, Home Décor, and Holidays and Events. Weddings are a huge category as well. If any of these categories are of interest, make sure to have a board (or two) dedicated to each of them.

A real-life example: Julie Hyzy pens the A White House Chef Mystery series with titles including *State of the Onion* and *Affairs of Steak*. Needless to say, Hyzy needs a Pinterest board dedicated to food. She may want to have a board dedicated to each ingredient that is mentioned in a book title, for instance one for steak recipes. If your book includes weddings, makes sure to have a board where you pin "dream wedding" material that might relate to your characters' nuptials.

For boards about home décor, let your fans get a peek at your personal style, without using photos of your own home. Jana has a board to showcase writer's cottages that reflect her personal style and it includes her own writer's retreat.

Your Books on Pinterest

Your books should have a board all to their own where you can pin covers and other book-related material. Many authors have separate boards for each book or series where pins may include actors that look like characters in the book, along with location photos and inspirational material they used to help write the story. If you use playlists to help stoke those creative juices, share those songs with your readers with Pinstamatic,[7] which helps you pin songs directly from Spotify.[8] Readers can then share those playlists with others or build one of their own.

Do you visit actual locations that become the settings for your books? If so, snap some decent photos, give them a home on your website, and then pin them to Pinterest. If someone clicks on the photo for the link because the image interested them, it should take them back to your website, where they will be exposed to more information about you and your books.

Books You Love

Many readers will have a board for either authors or books that they adore, and you will want to do the same. Here is a chance for you to show some love to your fellow author buds, along with those authors that inspire you.

Secret Boards

You can create a board that only you can see. A secret board is a great place to store pictures you are using for inspiration for your work-in-progress. Once the marketing efforts have started for the book, you can pin your "secret pins" to a public board about the book. You can also invite others to pin on a secret board. Both of you will see the secret board you shared on your Pinterest dashboard.

A Community Board

Pinterest allows you to invite others to post on your boards. While we would not suggest opening up all your boards to other pinners' input, we do suggest having a reader/fan community board. These boards will not only appear on your board list but also on their boards. Let's talk about two potentially beneficial ways to utilize this function.

The simplest way is to create a public board about your books and invite select readers to post items connected to the book on that board. These readers could be ones that have pinned about your books or shared your content. We'll talk more about to find them later. Not only will these readers feel special when you invite them, but they will also become a brand advocate for you, helping to spread the word about your books.

Another more complicated way is to have a secret team board to which you invite members of your street or marketing team. When it is time for launch, you can ask your street team to repin the pins from the "secret street team" board to their own public boards. You could do the same thing with the marketing pros from your publishing house, if they are heavily invested in Pinterest.

Unique Interests

Most of us have interests that are unique—or at least different from the top four Pinterest board categories. Jana has boards for Favorite Pubs and Windows to the Soul because she enjoys both a good night at a pub and beautiful stained glass windows. Tyra has boards about Cats, Dogs & Other Animals, Purses and Bags, and Geeky Stuff, reflecting her love of her fur babies, handbags and her gadget-geek personality. What board can you create that says something about you?

Once you've set up your boards, it is time to start pinning. Our website has a graphical illustration of how to pin, but let's go through a few basics.

Pinning

There are three basic ways to pin on Pinterest. The easiest way is to click on a "Pin It" button on a website. Sometimes it might just be the "P" part of the logo in a red circle, other times it will be a rectangular Pin It box, which is similar to Facebook and Twitter share buttons.

Secondly, you can use a Pin It button that installs into your web browser, making it easy to share content from any webpage regardless of whether the site has a Pinterest share button. Links to add a button to your browser can be found here.[9]

Lastly, you can upload a photo to Pinterest but it requires a few extra steps to link it to your website. After you upload and pin the photo, click on it from your board. You will now see your pin on the full screen and there will be an "edit" button at the top.

Click the button and there will be a field labeled "source." Input your website URL and "save." Now when people click on the photo they will be taken directly to your website. To help make images easily pinnable, create a special page for them on your site, which will be their permanent home.

Speaking of your website, include a Pinterest widget on your site so that your content, particularly your book covers, is easily pinnable. Pinterest has several available for usage on your website, depending on what you are trying to accomplish. You can find out more information here.[10]

After you have chosen what to pin, the next important step is the description. This is where you need to think about keywords and search engine optimization (SEO). If it is a pin relating to your book, don't forget to include the title and series name (if applicable) in your description, along with other keywords related to the genre.

Also, consider if it is necessary to post your pin to other social media platforms. Within the Pinterest pin interface, you can choose to share on Facebook and Twitter. We suggest you do this with extreme caution. Not only do you not want to clog up your followers' and fans' newsfeeds, the content that is shared on Facebook or Twitter should be different (or presented differently) than what is on Pinterest.

Repinning

Repinning is a great way to get content for your boards. By looking at the pins in your newsfeed and repinning the ones you are interested in, you can increase the content of your boards. You can also search for topics you are interested in for content to add as well. When you do repin someone's pin, don't forget to change the description if needed.

Blogs and Pinterest

Hopefully, by now you aren't wondering why we're talking about blogs and Pinterest together. It's simple: As mentioned earlier, over half (54%) of click-throughs on Pinterest pins go to blog posts. That means your blog posts have a good chance of being clicked on as well.

To ensure your posts make it to not only your Pinterest feed, but your readers, you need a Pin It button associated with *every* blog post. For most, it will be as simple as clicking a button or installing an app to tell your blog software or host that you want people to be able to share your post through Pinterest, just like they can share it on Facebook and Twitter.

If you want your blog posts pinnable to Pinterest, then you

have to include some type of graphic or photo in each and every post. This is a given—otherwise the visual element that is the defining feature of Pinterest is absent.

Though we did not include one under our discussion about boards, if you are a consistent blogger you may want to create a board entitled "My Blog Posts? where you can pin a picture from each post when it is launched. One word of caution here: make sure you are pinning from the page with only the post you want to pin, and not from the front page of your blog. It's the difference between standing outside a bookstore or walking up to the specific section designated for romance novels. That way your readers don't have to scroll through to find the blog entry, which might be buried quite deeply over time.

Pinterest Analytics

If you have a business account, Pinterest web analytics allow you to see how your pins are performing and which pins and boards are getting you the most impressions, clicks and repins. Under the dropdown menu from your user name (in the upper right-hand corner), you can click on "analytics" to view your Pinterest page's statistics.

You can easily look at metrics for a seven, fourteen or thirty-day timespan, or simply click on the dates and select the specific time period to judge your performance. The first tab will summarize your "Site Metrics."

The first row of site metrics tells you the number of pins and pinners that shared content from your registered website to Pinterest (and to their individual boards). The second section tells you the number of repins and repinners. Next, Pinterest gives you an idea about your impressions and reach.

The "Impression" number is the average number of times your pins appeared in someone's feed, in search results, or on boards. Reach tells you the average number of users who saw your pins. Impressions are often greater than reach because a single pinner may view more than one of your pins. Lastly you'll see the average number of daily clicks to your website originating *from* Pinterest and the average number of visitors *to* your site from Pinterest.

The "Most Recent" tab showcases your most recent pins. The third tab is for highlighting your "Most Repinned" posts, followed by the "Most Clicked." Both of these sections are good ways to determine what content is resonating with your audience.

Finding Influencers on Pinterest

Back under our discussion of boards, we mentioned creating a private one for members of your street or marketing team. Some of you will already have a street team ready to go and spread the gospel about all things related to your writing. If you are just starting your career, you may have no one besides friends and family to help you out. Pinterest can help you discover readers who like your genre or more specifically, your books.

The search box in the upper left-hand corner of the screen (to the left of the word Pinterest at the top of the page) is one of your primary tools for finding possible fans. You can search for pins, boards or pinners, but for this exercise, we're going to focus on pins and boards. Do a search in both of these for your name, your book title and your series title.

As you scan over the pins, are you seeing a couple of names consistently, a few pinners who have pinned content related to you or your books multiple times? Go check out these pinners' boards and follow some of them. You may even want to consider inviting them to the public fans board or even your private street team board.

If someone has posted a photo of meeting you, and you remember meeting them as well, consider commenting on their pin. If you meant enough to someone for them to post such a photo, a comment will strengthen that emotional connection between you and your fan. This is also true of a pinner who has named a board after you, your book or series. This fan is a highly engaged and certainly someone you should consider inviting to pin on your public board and be a member of your street team.

If you have done guest blogging or online interviews, you may find pins for them when you search. If you have not pinned these on your boards, you will want to do so. This may also be a good time to comment as well.

Next, you want to search pins and boards for your genre(s). This can help you spot cover trends and discover what readers are hot to read. It can give you a snapshot of the industry from the here and now.

Tips and Tricks

And now, four Pinterest tips and tricks.

First up is "Pin Alerts." Pin Alerts is currently in beta and the website (and name) may change.[11] At their site, you sign up to receive e-mails when someone pins content from your website. If you have a highly trafficked or pinned site, you may want to consider setting up a separate e-mail address just to receive alerts like this one so they do not clog your personal or work e-mails.

Repinned is a site that keeps track of the most pinned topics and pins, as well as information on which boards and pinners have the most followers.[12]

You may want to check out some of the top followers like Joy Cho, who is a blogger with over thirteen million fans. Yes, you read that right. *Million*. You can also see which pins have been repinned and liked the most. You may want to add them to your boards or you may want to see what the general trends are and pin accordingly. If you are logged in to Pinterest, you can pin directly from the repinned.net site.

Next up are two content creation tools—Pinstamatic and PicMonkey. Pinstamatic will help you create a pin of your website, a quote pin, link a Spotify song or create a calendar post—great to highlight upcoming appearances. PicMonkey will help you edit and touch up photos as well, as design quote pictures and collages.[13]

Hopefully, we have inspired you to make your website and blog Pinterest-friendly, and to start filling up your own boards with content we can't wait to repin.

Chapter Eight Endnotes/Links

1. Ahalogy 2014 Pinterest Media Consumption Study— (http://www.ahalogy.com/research/)
2. Social Network Demographics— (http://mashable.com/2013/12/30/social-network-demographics-2013/) and (http://www.pewinternet.org/files/2013/12/PIP_Social-Networking-2013.pdf
3. Pinterest Discovery— (paid site) (http://www.ft.com/cms/s/0/482bbd18-ed16-11e3-8963-00144feabdc0.html#slide0)
4. Ahalogy 2014 Pinterest Media Consumption Study— (http://www.ahalogy.com/research/)
5. Snapshot of Reading in America— (http://www.pewinternet.org/2014/01/16/a-snapshot-of-reading-in-america-in-2013/)
6. Trends in Consumer Book Buying— (http://randomnotes.randomhouse.com/trends-in-consumer-book-buying-infographic/)
7. Pinstamatic—(http://pinstamatic.com/)
8. Spotify—(https://www.spotify.com/us/)
9. Link to add Pinterest button to your browser— (http://about.pinterest.com/en/goodies)
10. Pin It Buttons— (http://business.pinterest.com/en/widget-builder#do_pin_it_button) and (http://business.pinterest.com/en/pin-it-button)
11. Pin Alerts—(http://www.pinalerts.com)
12. Repinned—(http://www.repinned.net)
13. PicMonkey—(http://www.picmonkey.com/)

9 Blogs: Journals of the Digital Age

Web logs (blogs) are everywhere.

For an author, blogging is a creative way to stay in touch with your readers, share a bit about yourself and generate interest in upcoming books.

Blogs can be ornate or simple and sparse. They are, in essence, a reflection of their creator. Some of you are world-class bloggers: your posts shine and perhaps even generate their own form of gravity, pulling everyone into your orbit. The majority of us? Not so much.

Like other forms of social media, if you're good with blogging—go for it. If not, don't feel guilty. There are other things you can do on a blog besides write entertaining posts about your latest traveling snafu. Figure out which method works for you and capitalize on it by using your blog to drive your fans (and potential readers) to your website.

One thing we discovered while writing this chapter was that we have different views about modern-day blogs. Though we both started off in the "old days" on a platform known as LiveJournal (more on that later), we have divergent ideas about what a "blog" means now.

Jana sees a blog as being separate from a person's website and containing mostly, if not all, long form posts. It's just another social media channel for her.

Whereas Tyra sees a blog as part of your website and a primary way for you to communicate with your fans, a site you control and that can be used to drive inbound traffic. Changes in blog platforms, particularly with the introduction of WordPress, have allowed both of these views to peacefully co-exist. You will need to figure out what works best for you.

The important point about blogging—and the reason we do encourage you to have at least some type of blog—is inbound traffic. If you remember in the section about websites, we mentioned that your site is the only place on the worldwide web where you have total control and, because of that, you need a consistent and creative way to drive people to your web home. Book covers and photos pinned to Pinterest can drive traffic. Tweets about deleted scenes from your latest book can work the same as a Facebook post about apple pie, especially when it includes a link to the recipe on your website.

Where can you put all these book covers, photos, deleted scenes and apple pie recipes? Why, a blog, of course! So, don't fret if you aren't the best at writing long-form blog posts. There are other things you can do besides wax poetic on a topic.

Creating a Blog

The first decision you have to make is whether you are going to host your blog on your website or have it reside on one of the numerous available platforms. If you choose to have your blog somewhere else other than your site, realize that though you will drive people to your blog, it will not drive people to your website. It also will not help with your website's search engine optimization.

There are several blogging publishing tools that also allow you to create a website around your blog. Again, you will need to decide what is best for your situation.

Starting a new blog has grown easier over the years because of an array of new publishing tools. If you're unsure what platform you want to use (WordPress and Goggle's Blogger being two of the most popular), ask your fellow authors what they use and if are they pleased with the results.[1][2]

If you're unsure which is best, perhaps conduct a test run on a couple of the formats to see which you like better, which you find easiest to manage, etc.

To get you started, check out The Next Web's "Top 15 Blogging and Publishing Platforms Today," which offers a review of each of the platforms' strengths and weaknesses.[3] TNW rated WordPress as the top blog site/platform. One of the confusing aspects of WordPress is that it is actually two things: WordPress.com and WordPress.org.

At WordPress.com, you can create a blog (blogname.wordpress.com) and use a wide array of templates, both free and paid. You can choose exactly how you want your blog to look. At WordPress.org, you download the open-source software and use it to create your own WordPress-based site.

WordPress allows you to schedule future posts—always a plus—and to quickly respond to comments from your readers. It has numerous "plug-ins" such as a calendar option or foreign translation (should your site draw international readers). As with any publishing host, there is a learning curve, but in the end WordPress will create a professional site. It may be worth hiring a professional to help you get up and running, but you should be able to update the site with only occasional assistance.

Curiously, LiveJournal, a blogger site that has been in existence since 1999, is the platform where both of us started. Jana moved to a WordPress site because her site did not require reader registration (which LJ does require). Tyra, on the other hand, moved to Blogger because of its integration with Google's other tools. However, many authors who began blogging on LiveJournal have remained there. It all depends on how you wish to manage your presence online.

Tumblr

Tumblr is another popular blogging platform that has been acquired by Yahoo![4] With a look at the site's demographics, we find that Tumblr users are slightly more female than male, and 27% of those users are 18-24. This is a young platform but also an educated one, with 57% of users having some college education. However, 68% earn $0-50K—so though educated (possibly in the midst of their education), they are not in an upper income segment. On Facebook or Pinterest, you are likely to find mothers, but two-thirds of Tumblr users are childless. If this younger audience lands in your target market, then you should be engaging your readers on Tumblr.

Because of its unique demographics, Tumblr is a different kind of blogging platform with a souped-up social aspect. It is quick and easy, driven by photos, memes and GIFs. Increased interaction with other Tumblr accounts is expected in the form of "hearting something" or reblogging (sharing it on your Tumblr wall).

There are accounts that are more of a mix of quick posts and longer form. One of the best examples is by Felicia Day.[5] Though not an author, she is an actress and scriptwriter and Tyra is one of her fangirls. Felicia uses Tumblr to post quick photos, cat GIFs, and links to her YouTube Channel (Geek and Sundry) videos, as well as traditional blogs.

Creating & Updating Your Blog

Here are some basic things to keep in mind while you're creating (or updating) your blog:
- Ensure that your blog's theme, graphics and images are consistent with your other social media sites

- Display prominent links to those other sites (website, Facebook, Twitter, etc.)
- Include an up-to-date mini biography and author photograph, if possible
- Add a "subscribe to this blog" button
- Use a contact form so readers can e-mail you
- If you have a newsletter, add an easy-to-use sign-up form
- Use Tags to separate your blogs into topics for convenient reader searches

Fantasy author Jim Hines' blog is an excellent example of a friendly and accessible blog, one that has a clean layout, frequent reader engagement and savvy author marketing.[6] It is built on the WordPress platform.

The cover image on his blog is the same one as on his website, and he has positioned the various social media icons at the very top, including the RSS (Real Simple Syndication) button. All of these are important, especially the RSS symbol, which will allow his readers to receive a notification via e-mail whenever he updates his blog. That saves them from having to check his site every few days.

Jim is a *prolific* blogger who posts nearly every day unless he's at a convention. Those posts are always of interest, whether they are about his books or concerning topical issues. As a former rape crisis counselor, Jim is very passionate about victims' rights, and doesn't shy away from tackling difficult subjects such as sexual harassment at conventions.

His Cool Stuff Friday posts are a collection of links to humorous and newsworthy articles.[7] When we were writing this book, his Cool Friday selection for that week included a link to pictures of children reading books to shelter cats. It pegged the adorable meter. Jim is one of those world-class bloggers we mentioned earlier, and it's earned him a loyal fandom.

Besides the intriguing content, Jim's blog page has an easy-to-navigate format that makes it easy for you to scroll though his posts or search for certain topics of interest. He has added his Twitter feed on the right, along with a "follow me" button. He also includes a list and links to his free fiction, which is a smart way to introduce new readers to his writing style. Once they've read one of his stories, they'll be more inclined to buy his books.

Urban Fantasy and Paranormal Romance authors, Ilona Andrews (actually a husband-and-wife team) are frequent bloggers, with topics ranging from the escapades of their two teen daughters (Kid 1 and Kid 2) to their pets.[8] They also talk about what it's like to be a two-author family, about their book series, share snippets from fellow authors, review online games and movies, and let us know what life is like in Texas.

In short, they're real people with real-life issues (like the time the dishwasher leaked and ruined their kitchen floor). This familiarity is what brings their readers back day after day. And again, they've tied the theme of their blog to their website, so their branding is consistent. In the right-hand column, they've included handy links to their publishing schedule, release schedule, snippets, writing articles and wallpapers (in case their readers want to post their gorgeous images on their own sites).[9]

Both Hines and the Andrews know how to give their visitors a constantly changing array of interesting articles and visuals. Make those your goals as well.

Content

Once you've developed your blog's theme, the next step is to decide on a writing style. Are you chatty, like you're talking to a friend over a decaf mocha latte? Or are you more serious? Decide up front what kind of style you're going to use and stick with it. Readers will expect that consistency.

Unfortunately, blog postings do not just write themselves. Unless you're blessed with adorable babies, quirky teens, photogenic cats, dogs, guinea pigs or pet lizards, dreaming up your next post can be tough.

Jana has always found it hard to write blog entries, which is one of the reasons she's more active on Facebook. She jokes that one reason she adopted a rescue cat is so she'd have something to blog about, and Dali has indeed provided the occasional post material. After a reader begged her to post something (after a three-month silence) she's tried very hard to do so, at least once per week. Still, she's not as prolific as other authors.

> **Pro Tip:** If you find an interesting article or post while you're surfing, write a brief post about it and add a link to that item. As long as you don't copy the content and claim it as your own, you're golden.

If you have a hobby such as photography or you collect pictures of cute guys (who wouldn't?), be sure to blog about that passion. Author Elizabeth May frequently shares her breathtaking pictures of Scotland.[10] Since she's a professional photographer in her own right and her Falconer series is set in Scotland, it's a very subtle way of marketing to her readers.

One of the most important aspects of any blog is keeping it

current, meaning you should post new and interesting content approximately two to four times (either weekly or bi-weekly). If you keep the site active, readers will return again and again to find out what's new in your world. Blogging gives you content to share on other social platforms—or on an inbound landing spot for certain posts. Some of the most successful bloggers have built an impressive fandom, not only because they're writing great books, but also because they're interacting with their readers on a regular basis.

Spam

Blogs attract your readers, but they are also a magnet for spammers. Blogs are a great target for people who want you to believe that your post is the most glorious they've ever read, providing you just click their link. A link that can either try to sell you stuff or download malware. Unless you find a way to block them, spammers can clog your blog's comments with tons of junk, making it hard for your genuine visitors to interact with you.

There are a number of ways to handle this spam. Some blog owners click on each post and "moderate" them—deciding if they're for real or if they're junk. Often you can set your blogger site so that you only need to moderate the first post by a new commenter, and the ones after that are automatically posted. WordPress maintains a blacklist of commenters and certain keywords to help filter the detritus.

You can install a plug-in on your blog to do the job for you. One of the co-authors of this book uses Akismet, which is spam-filtering software.[11] Unlike some e-mail filters, Akismet does a good job. With various plans, you can tailor how much spam protection you need.

Your contact form will also generate spam, and the best way to handle that is to install a Captcha plug-in for your site.[12] As they say, their job is "telling humans and computers apart automatically." The Captcha process involves typing in specific information (usually words and numbers) that computer robots cannot duplicate. It's also sometimes difficult for your readers and may require two or three tries.

Captcha isn't just for your contact form; it can also prevent spammers from "scraping" e-mail addresses from your site. You can obscure your address(es) from spammers by placing that information inside a JPEG image.

Blog Tours

As fewer and fewer authors are being sent on book tours and bookstores have been reluctant to host book signings, online tours have begun to fill that gap. Blog tours are usually scheduled around a new book release and are designed so an author can "stop by" a new blog each day. Usually these are top reviewers' blogs that have a significant number of followers, though sometimes that is not the case. It's nice to share the love with the newer reviewers as well.

Planning a blog tour can be time consuming, but fortunately there are professionals available to handle those logistics for you. Jana uses a company out of England to set up her blog tours, both for her books in the U.S. and the U.K. *Dark World Books* contacts the bloggers, sets up the terms of the tour and determines what those blog hosts expect in return.[13] They design the advertising banners and promote the tour for you.

Often the blog hosts want you to answer a series of questions or write a special post for your 'day' on their site, usually tied to themes in the new book. When you're preparing for a week-long or ten-day blog tour, that can be a lot of content to write up front. Usually there are giveaways of some sort to encourage readers to comment. For example, asking them to "Like" your Facebook page or follow you on Twitter in order to be eligible for the giveaways.

It is expected that you and the blog host will promote your visit, and you should expect to spend time on their site during your "day." along with a couple of days afterward so you can chat with the blog's visitors as they comment.

The great part? You can conduct a blog tour in your workout clothes and zombie bunny slippers while working on your next manuscript, all from home. No TSA security lines at the airport, no jet lag, no hauling your luggage all over the place. It's a high-tech way to reach your readers with minimal wear and tear on your health and productivity.

Ah, but are blog tours effective? It all depends on whom you ask. A tour is a good (and inexpensive) way to generate buzz the week of your book launch, and they do connect you to your target market, providing you've done your research and are posting on influencers' blogs. Will that result in increased book sales? Unknown. A thoughtful discussion on this very topic at Passive Voice might help you form an opinion.[14] Be sure to check out the comments beneath the post.

Blogs are your very own piece of real estate, and you should use that landscape to your best advantage. Entertain and inform your readers, share your passions, your books, your life as a author. In return, your readers will share your triumphs, commiserate with you when life isn't fair and eagerly look forward to your next book.

Chapter Nine Endnotes/Links

1. WordPress—(http://wordpress.com/website/?source=google&campaign=hsb&gclid=CImY9L6egcACFeRj7AodiE8Aww)
2. Blogger—(https://www.blogger.com/features)
3. The Next Web Blog Reviews—(http://thenextweb.com/apps/2013/08/16/best-blogging-services/)
4. Tumblr—(http://unwrapping.tumblr.com/post/43854304176/tumblr-demographics)
5. Felicia Day—(http://thisfeliciaday.tumblr.com)
6. Jim Hines' blog—(http://www.jimchines.com/blog/)
7. Jim Hines' Cool Stuff Friday—(http://www.jimchines.com/2014/08/cool-stuff-friday-34/)
8. Ilona Andrews' blog—(http://www.ilona-andrews.com/blog/)
9. Ilona Andrews' Wallpaper Images—(http://www.ilona-andrews.com/blog/category/wallpapers/)
10. Elizabeth May Photography—(http://www.elizabethmaywrites.com/updates/)
11. Akismet—(http://akismet.com/)
12. Captcha—(http://www.captcha.net/)
13. Dark World Books—(http://www.darkworldbooks.com/)
14. Passive Voice on Blog Tours—(http://www.thepassivevoice.com/03/2013/planning-a-blog-tour-think-twice/)

10 Goodreads: Social Media for Readers

"Meet your next favorite book."

That's Goodreads' promise, and with over thirty million members and nine hundred *million* books in their catalog, they can deliver.[1] If you are going to be involved with one book-related social media site, make it Goodreads because of its sheer size.

Launched in 2007 as an online community for readers, Goodreads grew large enough to attract Amazon's attention, who bought them in 2013. Their competitors include LibraryThing (1800 members) and Shelfari, both of whom have considerably smaller memberships. Amazon owns helfari and they have a stake in LibraryThing as well. Given that they're the world's biggest online book retailer, that interest makes financial sense.

It's All About the Readers

Goodreads' biggest strength is its reader community. Members post reviews, rate books and comment on other members' reviews. They can build lists of their favorite reads, such as Best Romance or Science Fiction of 2013. They can also join communities of like-minded readers and enjoy extras such as trivia, quizzes and quotes.

Based on your reading history, and using twenty *billion* data points, Goodreads will even make recommendations to match you with the ideal book. The sheer analytical power of this site, coupled with the enthusiastic interaction of its members, and the power of their recommendations, makes Goodreads a prime location for author promotion. But first, you really need to invest in the reader side of the equation.

Signing up is easy—enter your name, e-mail and password (or signup using your Facebook, Twitter, Google or Amazon accounts). Once you have an account, you can set up a virtual "bookshelf." Then it's only a matter of adding the books you're currently reading, have read or hope to read.

Since Goodreads is all about building community, they'll request access to your friends from Facebook, Gmail, Yahoo! and Twitter. If you agree, they'll compare those names and e-mails with their database and determine who is already on Goodreads so you can connect with them. If you wish, you can skip this step.

So that they can get an idea of what you like to read, GR asks you to select your favorite genres from among forty different categories, including romance, mystery, Young Adult, non-fiction, history, etc.

Then it asks you to rate the books you've read. You can either select from the books whose cover images they've displayed, or you can search for them on your own. Once you hit twenty ratings, GR will begin recommending reads based on your likes/dislikes.

Because they're tied to Amazon, there's a link at the top right of the recommendations page that allows you to link Amazon book purchases to your GR account. You can then choose which of the purchased books you wish to shelve and rate. Once you're through registration, Goodreads welcomes you to their site and shows you current messages from the community.

Navigating Around Goodreads

Navigating around the site is simple via the links at the top of the page, but trust us, there's a *lot* to check out. To start with, click the "My Books" link and you can see the entire list of every book you've rated. (You can also display this list by cover image.)

Each book lists the average GR rating, your rating, your review (if you've posted one), the date you read the book and when you posted it to your bookshelf. If you click on one of those books, you'll see the complete review, as well as any comments by other members. If this book is in your To Be Read pile, that can be designated as well.

To make it easy to share your reviews, in a box on the right side of the page is an HTML version you can plug right into your blog. You can also recommend the book on Google+ or "Like" it on Facebook.

This page allows customization of the various columns by ISBN, whether you own the book, etc. With the "Print" option you can create a portable book list for when you're making the rounds of the bookstores, solving that age-old question, "Do I have this one at home?"

Jana's stats show that she's read 475+ books and has a few in her TBR pile. She's not always been good about posting the books she's read (she usually devours at least 100-150 per year, so that number is lower than it should be). Also, she hasn't broken those entries into additional bookshelves, but she does have that option.

Based on Jana's reading history, Goodreads suggests books in the following genres: contemporary, crime, paranormal and Chick Lit. Oddly, that last category isn't spot on, since she doesn't usually read that sub-genre. Goodreads' suggestions in that category include Historical Romance and Romantic Suspense (which definitely aren't Chick Lit). Still, there were a number of books they suggested she would check out.

At the top of the recommendations page is a Sponsored Recommendation—on this particular day it's Diana Gabaldon's *Outlander,* just in time for the Starz television show's debut. Considering GR's huge membership, a sponsored post like this will catch quite a few eyes.

Clicking the "Groups" link gives you what seems like a zillion options for communities that might fit your reading preferences. You should spend time checking these out, as groups may be managed by readers/reviewers who hold great influence on the followers of that community. Post in the appropriate groups, talking about books you love. Resist the urge to chat only about your book(s), as that's not well received.

You can also create your very own group, which is a great way to establish yourself as an influencer on GR.

Returning to the main book list page, along the left side is another set of tools, including the ever-popular "Recommendations" and "Add Amazon Book Purchases." Of interest to us is the "Widgets" option. This is a license to have some creative fun!

Because Goodreads wants their members to spread their name far and wide, they've created widgets to make that a simple task. Using one of the widgets, you can showcase the contents of your bookshelf on any JavaScript-friendly site. Just click which books you want included and GR will generate the HTML, which you can plug into your blog or website. You can also create a GR "Updates" widget, a cover image grid or quote widget, a "'Shelf Badge." an "E-Mail Footer" or "Reading Challenge" widget.

All of these actions are clever ways to show your love for books and offer additional content for your Facebook page, blog or website. When you sign up for GR Author Pages, these widgets will point back to your GR profile and help build reader interest.

Goodreads for the Author

Once you're comfortable puttering around the site, have read a number of reviews to get a sense of GR's "tone," made new friends and commented on their reviews, it's time to dig into the author side of Goodreads. No matter whether you're writing fiction or non-fiction, GR has something for everyone.[2]

Where Facebook and Twitter are invaluable when it comes to real-time interaction, Goodreads is all about books. The people are here to form social connections with other bibliophiles and to find the next great read. It's your job to help them realize it might be one of your books.

The best way to do that is to become part of Goodreads' Author Program. A handy list of "Frequently Asked Questions" (FAQs) is available on their site.[3]

They do not restrict authors by geography (meaning you can live outside the United States) or by which route (traditional publishing vs. independently published).

If your books aren't listed on the site, you can manually add the book yourself, providing you have verified your e-mail address so GR knows you're for "real." If your book(s) are already listed on the site, click on your name and it will take you to your author profile, which is separate to your personal one for now. If at any time you're confused, consult the Goodreads librarians, the true wizards behind the site. These fine people volunteer their time to keep things orderly and are a great resource.[4]

Once you reach this profile page, scroll to the bottom and click the "Is this you?" link, which will send a request to become one of GR's authors. It takes a few business days for the request to go through, and then you'll receive an e-mail verifying that acceptance, along with further instructions. *At this point your personal and author profiles have been combined.*

A word of caution: the reviews that you posted as a reader are now associated with you *as an author,* and any future reviews you make will be as well.

Once you're official, flesh out that profile, remembering this might be the first point of contact for a new reader. Include a top-quality photograph, a biography and all the social media links so the readers can find you elsewhere on the web.

You have the option to link your blog posts directly to Goodreads, which we recommend. That way you don't have to remember to copy the latest post to their site. Do drop by every few days to respond to comments and messages and to accept friend requests. Because you'll be getting them.

The author profile page also shows you how many friends and fans you have. These numbers will not be the same, as not all fans are going to request to become your friend. Below the list of your books are messages from your friends and a Topics section where your name has been mentioned. It's good to keep an eye on this so you respond to your readers' comments as the need arises.

Making the Most of Goodreads

Using GR's reach to its full potential means you have to post content the members will enjoy. So here's a few ideas to help you out, some of which are often used on other social media sites:

Videos—Upload that latest book trailer or a current interview

Quotes—Post your favorite quotes from your books

Snippets or Outtakes—Readers love seeing scenes that didn't make the final cut, or teaser scenes for the newest novel

Quizzes—Write a quiz based on your book or characters

Widgets—Don't forget those nifty widgets to help bring readers to your GR page

Giveaways—Goodreads will host your book giveaway, which always stokes reader interest[5]

(Note—GR's rules do not allow you to use the readers' contact information for anything other than mailing them your book, so do not add them to your newsletter lists.)

Winners are encouraged, but not required, to submit reviews. Also, as GR warns that you are responsible for shipment of the books. You might consider offering two giveaways—one right before the new book launches and then one a month or two down the line, to generate additional interest for those who missed the first giveaway.

Events—List your upcoming events here, including book signings, talks and other appearances. You can invite friends to the event as well.

Ask the Author

Goodreads recently added a new feature entitled Ask the Author, beginning with fifty-four hand-selected interviewees (with big names like Kathy Reichs, Douglas Preston and Holly Black.) In time, the program will expand to include all GR's authors (some one hundred thousand souls).[6] According to Patrick Brown, director of author marketing, the program is designed to help deepen the connection between readers and authors, saving the latter from having to answer the same questions over and over.[7]

The feature is restricted to office hours to allow the author some respite, and the questions will not appear on the site until they've been answered and posted. You can turn off Ask the Author when you wish for quiet time or when you're at an appearance or book signing.

Once Goodreads opens this option up to all authors, we feel this will be an excellent way to keep in touch with your fans and help build on that relationship.

The Dark Side of Goodreads

Substantial membership numbers can generate Olympic-level drama and Goodreads has had its share of dust-ups. Since this a reader-driven site, book reviews can be wonderfully uplifting and helpful (both for the reader and the author) or immensely painful. Much like on Amazon, these reviewers do not tend to pull their punches.

Because of this, some authors will not read their GR reviews, for fear their creativity may be affected and disrupt their daily word count. Even those of us with the toughest skins may find some of these reviews hard to handle. Many authors choose to play on the outer edges of the sandbox, resisting the urge to scroll down to see what the reviewers think of their latest effort.

In their defense, reviewers feel they have the right to rate and review books as they see fit. Sometimes that freedom spills over into direct and cutting remarks regarding the author. Depending on the reviewer, she may believe that it isn't her job to offer constructive criticism geared toward the author, that her focus should solely be the reader. Other reviewers are keen to offer insights that benefit both the author and other readers. It all depends on the person behind the review.

Authors, for the most part, hope that the reviews would confine themselves to the book itself, not straying into personal attacks or innuendo, but sometimes they do. Personal vendettas may play a part, so much so that at times it seems that reviewers and authors are from two different planets.

GR boasts of thirty-four *million* reviews, and you can bet a number of those aren't pleasant to read, especially if you were the author of one of the books that was panned. Bad reviews are part of the publishing landscape, but how do you respond when someone takes a review too far? If you complain, either on the Goodreads site or on your blog—which may be streamed onto the GR site—fireworks can commence. We're talking World Wide Web-sized fireworks.

Former literary agent (and author) Nathan Bransford tackled this subject in a blog posted entitled "The Bullies of Goodreads."[8]

As the issue of author bullying became recognized, antibullying groups formed. Unfortunately, not all of them stayed on the straight and narrow, allegedly using unprofessional tactics to bully the reviewers. No matter who did what on either side of this controversy, there were no winners once tempers flared.

Jana has commented on negative/inaccurate Goodreads reviews on only two occasions, fully aware she was holding the Internet equivalent of a live grenade. She considers herself lucky that the situation did not escalate. That being said, neither of the authors of this book recommends such a tactic if you can avoid it. Make a misstep and you're inviting the situation to go viral, and not in a good way.

If you are physically threatened, or the reviewer has resorted to libelous statements (charges of plagiarism, for instance), contact Goodreads and let them handle it. In the long run, you'll be better off that way because these verbal firestorms can escalate dramatically, and they never stay just on one site. Always remember that you, the author, have the most to lose in these battles.

On the plus side, there is nothing wrong with posting a "Thanks for the great review!" to acknowledge someone's efforts on your behalf.

Despite the issues between authors and reviewers, Goodreads remains one of the best sites to connect with readers on a one-to-one basis. Spending time learning how Goodreads works, interacting with other readers and reviewers and offering content in return (reviews, fleshing out your book list, commenting on other books' reviews) is a solid way to get you noticed in the community.

Chapter Ten Endnotes/Links

1. Goodreads—(https://www.goodreads.com/about/us)
2. Goodreads Author Program—(https://www.Goodreads.com/author/program)
3. Goodreads' Author FAQs—(https://www.goodreads.com/help/list/author_program)
4. Goodreads Librarians—(https://www.Goodreads.com/group/show/220.Goodreads_Librarians)
5. Goodreads Giveaways— (registration required) (https://www.Goodreads.com/giveaway/new)
6. Ask the Author—(https://www.Goodreads.com/featured_lists/117-featured-authors-answering-questions)
7. Ask the Author—(http://www.washingtonpost.com/blogs/style-blog/wp/2014/05/21/Goodreads-wants-you-to-ask-the-author/)
8. Nathan Bransford The Bullies of Goodreads—(http://blog.nathanbransford.com/2013/09/the-bullies-of-goodreads.html)

11 Amazon's Author Central

We are all aware of Amazon's T-Rex-size footprint on the web, so why not take advantage of that imprint to promote your books? If you're like us, checking out an author's background information is just one of the steps we take while we're searching for a book on their site.

Amazon has made that easy with Author Central. Employing many of the same concepts as Goodreads, it's designed to be a one-stop location where you can learn more about a particular author and her books. Setting up your Author Central account is easy—Amazon gives you exact instructions.[1]

Do be aware that your book must be sold on their site (which is only fair) and it may take up to seven days before they can verify you are legitimately an author. One of those verification steps may involve contacting your publisher so they can be sure you're for real.

In the meantime, you should fill out your author profile. Start by adding your biography, which won't be seen by Amazon's customers until verification is complete. You can also add up to eight photographs and switch them out as needed. Other features, such as linking to your blog, are only available after Amazon has given you the thumbs up. If you're a co-author, they require a separate page for each of you.

You should fill out the information on the author profile as completely as possible, taking the time to ensure that the "image" you present on Amazon is in sync with your other social media platforms.

To help your readers connect to your author page without having to dig around for it, Amazon allows you to create a direct link you can add to the signature line on your e-mails and on your blog.[2]

It is our recommendation that you sync your blog with Amazon (much like you did with Goodreads). Set it and forget it. If you're on Twitter you can also set your tweets to appear in the section reserved for your feed (which is separate from the blog). That way, anything new you post will automatically be added there.

If you have video interviews or a book trailer, put those up on the page as well. If your fans have made videos relating to your books or series, obtain their permission to share them on the page. Entertaining content invites the reader further into your world, offering insights about you and your books.

Though you may be posting your upcoming events on Facebook or your blog, Amazon also gives you a chance to share them on their site.

And finally, your author page hosts your very own forum, where you and your readers can chat back and forth.

Managing Your Books

It's to Amazon's advantage that their customers can easily locate all your books, and the way to make that happen is for you to "claim" them.

By clicking "Add More Books" you can find your books on their site and add them to your bibliography. You can suggest corrections if the book's data is incorrect and let Amazon know if you're not listed as the author of a particular book.

They will also work with you to ensure the correct edition of your book is the one their visitors will find when they perform a search. You can also remove a book from your bibliography, should it be incorrectly attributed to you. Be aware that out-of-print editions are not removed. Once a book is on the site, it's there forever.

As an added bonus, Amazon offers a "Look Inside" feature that allows customers to read the first few pages of your book. This is specific to the edition, so the preview for your print book will look different than the one for your e-book.

To use this feature, you must submit an application and own the (regional) merchandizing rights to the book (meaning you have the right to distribute this edition in this part of the world). Along with the application you would supply Amazon with the digital files required for them to add this feature to your book's sales page.

If you are a traditionally published author, your publisher would be responsible for submitting this information. If you are an indie author, this ball falls in your court. If you're using Kindle as your e-book format and CreateSpace as your print service provider, this information flows to the Look Inside feature automatically.

Speaking from a readers' perspective, we both use this feature because it's a good way to get a sense of the author's writing style. Of then this is the last step before we click that "Buy" button.

Finding Your Amazon Readers

Once your author page is established, Amazon helps you find your readers in the "Customers Also Bought Items By" section on the lower right-hand portion of the page. These authors are who your customers purchased in addition to your books, your "competition" as it were.

In Jana's case, these authors include Jennifer L. Armentrout, C.C. Hunter, Sarah J. Maas and Richelle Mead. Knowing that her readers like these authors helps her target her posts, her advertising and her brand to better attract their interest.

No matter your feelings on Amazon and their relationship with authors and publishers, their site is a powerhouse and you should take every advantage of that ability to get books into the hands of their customers. As part of an ongoing social media strategy, an Amazon author page is well worth the time to create and maintain that presence.

Chapter Eleven Endnotes/Links

1. Amazon Author Central— (https://authorcentral.amazon.com/gp/help?topicID=200620850)
2. Personalized Amazon Author URL— (https://authorcentral.amazon.com/gp/help?ie=UTF8&topicID=200799660)

Congratulations!

You just finished *Socially Engaged* and lived to tell about it. Thank you for allowing us the opportunity to bend your ear about social media.

No one book can possibly cover all aspects of this dynamic medium, but hopefully we've given you an overview of the ever-changing social media landscape and how to best utilize various sites and their features to your advantage.

Please feel free to contact us with comments and questions. We'd love to hear from you either via our website, on our Facebook page or on Twitter.

Now go forth and socialize!!

Cheers,

Tyra & Jana

www.SocialMediaMuses.com
www.Facebook.com/SocialMuses
@SocialMuses

Index

A

Abney Park 50
Advertising 88
Affairs of Steak 96
Akismet 113
Amazon 12
Andrews, Ilona 60, 111
Android 88
ARC (advance reader copy) 51
Armentrout, Jennifer L. 83, 130
Author Brand 1
Author Central 128
Author Platform 1
Avon Books 55

B

Barnes & Noble 12
Bastille Day 75
Benson, Buster 89
Bing 41
Black, E. B. 55
Black, Holly 123
Blacklist 113
Blog 106
Blogger 40, 107
Blog Tour 114
Blume, Judy 4
Boards 92
Bowker Marketing Research 94
Brand Advocate 50
Brandi 55
Bronx Zoo's Cobra 78
Brown, Patrick 123

C

Captcha 114
Cast, Kristin 65
Cast, P.C 65
Cho, Joy 104
Circles 44
Clicks 73
Columbia Univ. Library 20
Communities 46
Content 36
Copyright 17
Copyright Advisory Office 20
Coverphotoz 21
CreateSpace 130
Creative Commons 21
Crusie, Jennifer 8

D

Dark-Hunter® 22
Dark World Books 114
Day, Felicia 109
Demographics 59
Demon Trappers® 23
DePaul University 25
Direct Message (DM) 80
DIY 22, 93
DragonCon 33

E

Engagement 69
Engagement Rate 74
Entangled Teen 83
European Union 19
Events 46
Evergreen Content 36
External Referrers 71
Extraordinary Contraptions 50

F

Facebook 9, 59
Facebook Analytics 69
Fanzine 53
FollowerWonk.com 55
For the Win (FTW) 82
FreeImages 21
Frenchy and the Punk 50
Frequently Asked Questions (FAQ) 121
Friedman, Jane 1
Friend 59

G

Gabaldon, Diana 120
Galbraith, Robert 3
Geek and Sundry 109
Geotagging 88
Gmail 40, 95
Godin, Seth 2
Goodreads 25, 117
Google 40
Google Alerts 38, 40
Google Analytics 35
Google Drive 40
Google+ (G+) 40, 95
Google Map 65
Grilo, Ana 53

H

Handle 78
Hangouts 45
Hashtag 54, 64, 80
HEA (Happy Ever After) 5
HFN (Happy For Now) 5
Hines, Jim 110
Hootsuite 87
Horror 50
House of Night 65
Hugo Award 53
Hunger Games 22
Hunter, C.C. 130
HyperText Markup Language 95, 119
Hyzy, Julie 96

I

Influencers 49

Insights 69
Instagram 88
Intellectual Property 17
iOS 88
Iowa State Fair 75
ISBN 119

J

Jackson, D.B. 84
Jackson, Percy 22
James, Thea 53
Jareo, Lori 18
Jones, Antony 50

K

Keywords 42
Kindle 130
Klout 52

L

Lecinski, Jim 24
LibraryThing 117
Likes 69
LinkedIn 88
LiveJournal 106
Logistics 80
Lucas, George 18

M

Malware 113
Mashable 86
Mass, Sarah J. 130
McCartney, Paul 4
McDonald's 5
Mead, Richelle 130
Morguefile 21, 37
MyKindaBook 87
My Pretty Pony 65

N

New Adult 49
New York Times 25

O

Objectives 33
Oreo 84
Organic Reach 67
Organic Treats 93
Outlander 120
Over 37
Overview 69

P

Page 59
Page Likes 69
Paid Reach 67
Passive Voice 115
People 69
Picasa 40
PicMonkey 104
Pin Alerts 103
Pinning 99
Pinstamatic 96, 104
Pinterest 9, 92
Pinterest Analytics 101
Pinterest Feed 93
Pinterest Widget 99
Pinwords 37
Pixlr 37
Post Reach 69
Posts 69
Potter, Harry 3
Preston, Douglas 123
Princeton University 25
Psy-Changeling 51

Q

Qualman, Eric 24
Quarter Pounder 5

R

Random House 94
Reach 69
Reading Challenge 120
Real Simple Syndication (RSS) 110
Recite This 21
Reichs, Kathy 123
Repinning 100
Retweet (RT) 80
Rice, Anne 16, 60
Robb, J.D. 2
Roberts, Nora 2
Romance Writers of America 83
Rowling, J.K. 3

S

Schmittauer, Amy 55
Search Engine Optimization 41, 99
SFBook.com 50
Shares 73
Shelfari 25, 117
Singh, Nalini 51
Sister Spooky 52
SMART 34
Smart Bitches 25, 50
Smith-Ready, Jeri 85
Socialnomics 24
Southwest Airlines 79
Spam 113
Spotify 97
Starbucks 50
Star Wars 18
Starz Television 120
State of the Onion 96
St. David's Day 75
Steampunk 50
Steampunk World's Fair 50
Steinem, Gloria 8
Street Team 51
SXSW 40

T

Taglines 8
Target Market 29
Team Logan 85
Team Zackary 85

Terms of Service 15
The Beatles 4
The New Yorker 25
The Shade Trilogy 85
Thurmeier, Heather 8
Timeline 59
Time Rovers® 23
TNW 108
To Be Read (TBR) 119
Top Influencer 53
Touchpoint 28
Trademark 17, 21
Trolls 14
Tumblr 109
Tweeter Karma 86
Twitonomy 89
Twitter 9, 78, 95

U

United Kingdom 19
URL 60
USA Today 25
U.S. Copyright Office 17

V

Vampire Chronicles 60
Viral 16, 126
Visits 69

W

Web Log 106
Website 9
White House Chef Mystery 96
Widgets 120
Wifey 4
William Shakespeare 79
Word of Mouth 23, 57
WordPress 106
World Cup 82
World of Mouth 24

Y

Yahoo! 41, 95
Yale University 25
Young Adult 74
YouTube 40, 55, 65

Z

Zero Moment of Truth 24

Tyra Burton

A senior lecturer of marketing at Kennesaw State University, Tyra is a gamer-girl, gadget geek at heart. An innovator in the classroom, she has taught students for over twenty years on topics ranging from Advertising to Mythology. She developed both the undergraduate and continuing education courses in social media marketing for KSU.

A native Georgian, she lives in Metro Atlanta with her husband and their three socially engaging fur babies.

www.TyraBurton.com

Jana Oliver

The winner of numerous awards, including the Maggie Award of Excellence, the Daphne du Maurier and the Prism Award, Jana Oliver's vivid imagination is never at rest. Her latest novel, *Briar Rose*, is a dark and twisted retelling of Sleeping Beauty set in the South.

When not wandering around the Internet or researching urban myths, Jana can be found in Atlanta with her very patient husband and a growing collection of single malt Scotch.

www.JanaOliver.com